Sociology of Interdisciplinarity

"This book is a welcome contribution to the energy social sciences field, based on broad consensus that sociological insights are crucial in addressing today's complex energy and climate challenges. Its novel multi-sited approach highlights practical case studies, strategies and methods to emphasise what works when it comes to interdisciplinarity and, importantly, what doesn't."

—Tessa Dunlop, Policy Analyst, *Directorate-General Joint Research Centre, European Commission*, Italy

"I highly welcome this important and timely contribution on the experiences of researching the energy transition. Through a critical analysis of empirical cases, it brings a much-needed social scientific understanding to the promises, complexities and challenges of doing interdisciplinary research, at a time when it has never been more essential to societies."

—Adel El Gammal, Secretary General, *European Energy Research Alliance*, Belgium

"This concise and timely book offers valuable concrete insights into how interdisciplinarity works in practice through an exploration of a variety of energy projects across Europe. In focussing not only on individual scholars and their experiences, but also on wider contexts such as the impacts of funding structures, different access to resources, and power relations, Sociology of Interdisciplinarity will be a significant resource for energy scholars and practitioners."

—Frances Fahy, Professor of Geography, *National University of Ireland Galway*, Ireland

"Silvast and Foulds show how unequal relationships across the disciplines contribute to how interdisciplinary collaborations are constructed. Although researchers in the technical disciplines routinely seek social scientists as partners, they also selectively construct social science to fit their existing methods and research programmes. This book shows how in many cases, the critical side of social sciences is replaced by a narrower focus on customers and demand. In doing so, Silvast and Foulds recuperate this critical side and bring it into the discussions of what interdisciplinary research can be."

—David Hess, *Professor of Sociology, Vanderbilt University*, USA

"Here is a refreshing perspective on interdisciplinarity. Drawing on energy research from Finland, Norway and the UK, Silvast and Foulds develop an analytical framework for understanding interdisciplinary research as a social activity. The framework promises to serve not only as an invaluable guide to thinking seriously about interdisciplinarity, but also as a reasoned antidote against uncritical hyperbole about its virtues."

—Janne I. Hukkinen, *Professor of Environmental Policy, University of Helsinki*, Finland

"Silvast and Foulds explore the practicalities behind building bridges between different approaches, canons and scientific cultures. This is good news to anyone who, like them, believes that breaking down such barriers and drawing on different sources of knowledge is essential for addressing today's societal challenges, not least the transition to climate-neutrality and a more sustainable energy system."

—Gerd Schönwälder, Policy Officer, *Directorate-General for Research and Innovation, European Commission*, Belgium

"This book provides living examples of how interdisciplinarity is constructed and negotiated through institutional rules and researchers' activities. A nuanced picture of various epistemic cultures serves to explore: how differently knowledges can be produced; how these differences can be recognised and understood; and how these knowledge can be exchanged between different communities. This thought-provoking research provides deeper insights into the practices of conducting interdisciplinary research. It is a must-have for all working interdisciplinarily and for those interested in knowledge production."

—Aleksandra Wagner, Associate Professor, *Institute of Sociology, Jagiellonian University*

"This book 'lifts the lid' on energy interdisciplinarity as practice. Drawing on case studies across a range of institutional and intellectual settings, and bringing these together with concepts from critical social science, the authors set out an ambitious new analytical framing. By recognising and discussing the challenges and ambivalences involved (especially for social scientists) alongside the opportunities, Silvast and Foulds offer a welcome and timely contribution."

—Mark Winskel, *Senior Lecturer in Science Technology & Innovation Studies, University of Edinburgh*, UK

"Even though I've been involved in interdisciplinary energy research for more than 20 years, this book is an eye opener for me. The authors offer refreshing insights into the project of doing interdisciplinary research, which will fascinate scholars. The nuanced discussion of empirical cases from energy research in the UK, Norway and Finland also makes the book equally relevant to practitioners and funders of interdisciplinary research."

—Tanja Winther, Professor and Head of FME *Include–Research centre for socially inclusive energy transition, University of Oslo*, Norway

Antti Silvast • Chris Foulds

Sociology of Interdisciplinarity

The Dynamics of Energy Research

Antti Silvast
Department of Interdisciplinary
Studies of Culture
NTNU
Trondheim, Norway

Chris Foulds
Global Sustainability Institute
Anglia Ruskin University
Cambridge, UK

ISBN 978-3-030-88454-3   ISBN 978-3-030-88455-0   (eBook)
https://doi.org/10.1007/978-3-030-88455-0

This Palgrave Macmillan imprint is published by the registered company Springer Nature Switzerland AG.
The registered company address is: Gewerbestrasse 11, 6330 Cham, Switzerland

For Molla and Tiina; Jenny, Zachary, and Thea

PREFACE

Interdisciplinarity is important. This message is so commonly stated—whether in, for example, call texts published by research funding organisations, university research and impact strategies, or by policymakers and practitioners—that the implications of what is actually being advocated for are often backgrounded or, at worse, forgotten. As such, the lived experiences of those delivering 'interdisciplinarity', and indeed the ripple effects onto what forms of knowledge are produced (and why), rarely get the attention that it deserves.

We contend that Science and Technology Studies (STS) has a critical role to play in exploring such issues. Much of STS was originally developed through studying the professional experiences of those in research and innovation settings (e.g. natural science laboratories), and thus it has numerous conceptual tools available to explain how and why different research and innovation actors do (not) collaborate 'successfully', and with what effects. We resolutely argue that these underutilised social scientific tools and resources therefore have much to offer the interdisciplinarity debate, which we demonstrate in this book via our studies of energy research projects.

Our research lives, as those of most of our colleagues, are strongly influenced by interdisciplinarity. These changed contexts are only partially defined by us, as researchers. They are shaped by an interplay of funding structures and university strategies; by an increasing number of academic disciplines that create novel combinations among them; and, not least, by global developments (e.g. climate change and digital transition) that create pressures to work in an interdisciplinary manner. This note on the scale is important because a considerable amount of what has been written about interdisciplinarity, including research that we cite in this book, draws only on personal experiences. It is true that such autoethnographic studies are important for shedding light on normally hidden dynamics and that reflections can be created only from immersing oneself within project contexts, and thus some of this book's chapters are indeed partially personal too. However, that said, and as the general motivation of the Social Sciences and Humanities (SSH) goes: these personal experiences are inadequate to fully understand the dynamics of interdisciplinarity. We need tools from the SSH to more fully understand the positions of those working in interdisciplinary projects. This book is an attempt to build such a toolbox for scholars and practitioners alike: as we show in this book, interdisciplinary projects can be analysed from within by using SSH tools themselves. SSH can therefore be used to investigate how SSH themselves are integrated or not with other disciplinary perspectives, going beyond isolated and limited descriptive reflections of interdisciplinarity that fail to sufficiently take advantage of the conceptual backing that SSH (and specifically STS) already offers.

Nevertheless, a personal note is necessary to frame our motivation. A desire to write this book has come from our own experiences of working in—and indeed being vocal advocates of—large-scale, consortia-based, interdisciplinary energy research projects. It has been through such projects that informal conversations with colleagues (from all manner of disciplinary backgrounds) have frequently raised concerns relating to, for example, frustrations of writing interdisciplinary project proposals that have to follow the 'scripts' of funding calls; what forms of knowledge carry what forms of authority within disciplinary integrations; various expectations of SSH from non-specialists; roles of quantitative (energy) models in organising interdisciplinary projects; common sources for disputes and controversies in ambitious interdisciplinary projects; and what disciplines in themselves actually offer. We felt that our common STS perspectives and recently collected data were directly able to respond to exactly these sorts of concerns. Indeed, we are somewhat surprised that so few of our STS colleagues have followed this path too.

We wrote the climax of this book—Chap. 5, on *A Sociology of Interdisciplinarity*—as an accessible introduction to relevant tools and resources from the STS literature. We specifically illustrated these by drawing on our three case studies (Chaps. 2, 3, 4); Chaps. 2 and 3 focused on interdisciplinary energy project's agendas and experiences, with Chap. 4 focusing on a more conventional monodisciplinary energy project and illustrated how the themes raised in the preceding two empirical chapters still remained pertinent. Given how Chap. 5 therefore represents this book's main contribution, it is obvious to see how it also directly led to the title of the book itself. In fact, it is exactly because of Chap. 5's critical role in assimilating this book's journey that we wish to make clear why we intentionally framed it as 'A Sociology of Interdisciplinarity', rather than 'The Sociology of Interdisciplinarity'. To us, this framing was critical to emphasise immediately that this book does not represent the definitive end-point; instead, we hope that this book represents the start of deeper discussions on how STS literatures may spark reflexive debate on the project organisation, professional experiences, academic cultures, and institutionalised environments sitting behind interdisciplinary pursuits and outcomes.

Our multi-scalar Sociology of Interdisciplinarity framework consists of six dimensions, and it is important for us to make clear that we hold a pluralist view on its application. As such, we do not believe the STS tools and resources (sitting behind each of the dimensions) are 'better' than any of the others; each dimension is appropriate for exploring particular issues. Indeed, the dimensions put the spotlight on usually overlooked aspects of interdisciplinary research project collaborations, and thus we would certainly hope for (and actively encourage) parallel consideration of the dimensions we raise, as well as further proposals for additional complementary dimensions too. What joins them altogether, though, is their shared common (sociotechnical) point of departure, which is what ensures they complement ontologically and epistemologically, rather than contradict, one another.

This book is certainly not intended only for STS scholars; nor is it only intended for those conducting (interdisciplinary) energy research. We have endeavoured to position this book at the boundaries between different knowledge-producing and knowledge-using communities, both inside and outside of energy research. Indeed, more broadly, we hope that colleagues may be more widely interested in our discussions around for how research and innovation systems are fundamentally organised, how this relates to normative goals of research, and what this all means for how SSH's expertise is integrated.

Trondheim, Norway   Antti Silvast
Cambridge, UK   Chris Foulds

ACKNOWLEDGEMENTS

We are grateful to numerous colleagues, without whom the book would not be what it is. Firstly, we thank all those who kindly gave up their time to be interviewed as part of generating Chaps. 2, 3, and 4's datasets. We acknowledge these data were collected via the following projects: the EPSRC National Centre for Energy Systems Integration (Grant number: EP/P001173/1; Chap. 2); the NTNU Energy Transition Initiative (Chap. 3); and the first author's doctoral dissertation (including the Academy of Finland project Managing Insecurities led by Turo-Kimmo Lehtonen, the Research Foundation of the University of Helsinki, the Kone Foundation, the Jenny and Antti Wihuri's Trust, and the Finnish Graduate School for Science, Technology and Innovation Studies; Chap. 4). Secondly, we recognise the valuable chapter review comments and insightful conversations from many of our colleagues, including Mark Winskel (University of Edinburgh), Simone Abram (Durham University), Marianne Ryghaug and Tomas Moe Skjølsvold (both NTNU), Sarah Royston and Rosie Robison (both Anglia Ruskin University), Sampsa Hyysalo (Aalto University), Salla Sariola, and Heta Tarkkala (both University of Helsinki). Thirdly, we note that many of the reflexive starting points for this book have come from our own experiences of interdisciplinary projects, and thus we acknowledge all our various collaborators in recent projects.

The authors' time on writing this book was primarily funded by NTNU Energy Transition Initiative (Silvast) and the Energy-SHIFTS project (Foulds), which was funded by the EU Horizon 2020 research and innovation programme under grant agreement number 826025.

The open access fees for this book were kindly paid by NTNU and Anglia Ruskin University's Open Access Fund.

CONTENTS

ABOUT THE AUTHORS

Chris Foulds is an associate professor at Anglia Ruskin University's Global Sustainability Institute. He was co-lead of the Horizon 2020 European Innovation Forum for energy-related Social Sciences and Humanities (Energy-SHIFTS, 2019–2021) and its predecessor EU platform (SHAPE ENERGY, 2017–2019). He employs a range of critical and reflexive social scientific perspectives from Science and Technology Studies, Sociology, and Human Geography, in investigating the relationships between policy and governance agendas, social practices and cultural conventions, and energy and sustainability transformations. He has a strong interest in unpicking the experiences and expectations associated with interdisciplinary pursuits. Foulds has led a number of projects funding through the EU, UK Research Councils, UK Government Departments, Newton, and various charities and foundations.

Antti Silvast holds a researcher position in the Norwegian University of Science and Technology (NTNU), Department of Interdisciplinary Studies of Culture. From December 2021, he will be an Associate Professor at the Technical University of Denmark (DTU), Department of Technology, Management and Economics, Innovation Division, the Responsible Technology Section. During his career, he has contributed to the forming of several successful European Horizon 2020 energy-related projects, including Checking Assumptions and Promoting Responsibility in Smart Development Projects (CANDID, at University of Edinburgh) and Sustainable Consumer Engagement and Demand Response

(SENDER, at NTNU). Since 2014, he has been an editor of *Science & Technology Studies*, the official journal of the European Association for the Study of Science and Technology. Silvast's research develops tools in STS to study multiple energy issues including interdisciplinary whole systems research, energy modelling, smart grids, control rooms, energy markets, and risk and resilience in energy systems. After defending his PhD in Helsinki (Finland), he held postdoctoral appointments at Princeton University (US), University of Edinburgh (UK), and Durham University (UK).

LIST OF FIGURES

LIST OF TABLES

CHAPTER 1

Introduction

Abstract This chapter provides background context on the calls for doing (more) interdisciplinarity and explains our own positioning as to what interdisciplinarity actually is, as well as what we believe this book contributes to the study of said interdisciplinarity. Specifically, we discuss mainstream arguments for why interdisciplinary research is deemed to be a worthwhile endeavour by many researchers, policymakers, funders, and so on. We build on this by arguing that there is a unique—and currently under-fulfilled—role to be played by Science and Technology Studies (STS) in exploring the sociological dimensions of how large-scale (energy) research projects are actually carried out. Alongside these wider landscape discussions, we explain what this book contributes to the study of interdisciplinarity and to energy research, through our empirics and STS-inspired ideas. We also make clear how we define interdisciplinarity and disciplines and explain how we focus on problem-focused research that may (or may not) involve external stakeholders.

Keywords Interdisciplinary • Knowledge integration • Problem-focused research • Disciplines • Sociotechnical

© The Author(s) 2022
A. Silvast, C. Foulds, *Sociology of Interdisciplinarity*,
https://doi.org/10.1007/978-3-030-88455-0_1

1.1   Introducing the Background Context

1.1.1   Interdisciplinary Research as a Mainstream Research Endeavour

Interdisciplinary research has been advocated as the zenith of research practice for many years, quite often in direct response to questions that cannot be answered (or even preliminarily investigated) by disciplines working separately (Jasanoff 2013). Indeed, common arguments for advocating interdisciplinarity often centre on fixing the 'poor connectivity' between disciplines, whether this implicitly/explicitly relates more to the knowledges or the knowledge-producing communities that map across such disciplinary classifications. The theory goes that interdisciplinarity fills knowledge gaps by improving disciplinary connectivity, thereby ensuring a "better *integration* of existing knowledge" (Hulme 2018, p. 333, emphasis in original). From there, claims that interdisciplinarity provides a more complete—perhaps even 'holistic' or 'whole systems'—perspective therefore often ensue.

Calls for doing interdisciplinary research have become so widespread and pervasive that doing and advocating for interdisciplinarity now very much occupies mainstream discourse—as shown by various contributions from researchers (e.g. Irwin et al. 2018; Nature 2015), educators (e.g. European University Association 2017; University of Essex 2020), funders (e.g. British Academy 2016; European Commission 2019; UKRI 2021), policy actors (e.g. HM Government 2017; Pellerin-Carlin et al. 2018), and related multi-stakeholder associations (e.g. Science Europe 2019) alike. Given this widespread multi-stakeholder agreement and its emergence "as a political preoccupation" (Barry and Born 2013, p. i), it is then no surprise that there have been calls for systemic, cultural changes that better enable the development and maintenance of interdisciplinarity (e.g. Caniglia et al. 2021).

Such is the widespread institutional support for interdisciplinarity that we believe scholars have become somewhat numb to the public support for interdisciplinarity. Essentially, explicit support and interest for interdisciplinarity is so commonplace that it has been rendered almost invisible or at least significantly backgrounded. Indeed, we would argue that vocal supporters of interdisciplinarity are rarely credited or congratulated—unlike they perhaps would have been 10–20 years ago—for endorsing or even directly funding interdisciplinary research. This is, of course, progress.

As part of this move towards greater interdisciplinarity in research and innovation, there have been explicit calls for interdisciplinary ambitions to account for the integration of Social Sciences and Humanities (SSH) approaches (Pedersen 2016). Indeed, this cause has been internationally championed at the broader SSH level by, for example, the European Alliance for Social Sciences and Humanities[1] (EASSH) and the Shaping Interdisciplinary Practices in Europe[2] (SHAPE-ID) EU Horizon 2020 project. Similarly, the EU platform for energy-related SSH[3] has also argued for SSH to be better integrated within the Framework Programmes of the EU (Foulds et al. 2020; Robison and Foulds 2019, 2021), as well as called for deeper reflections as to the roles afforded to SSH in interdisciplinary research, including considering its implications for the policy advice being generated (Foulds and Robison 2018; Royston and Foulds 2021). Such calls are built on the foundations of a range of works that demonstrate the underutilisation of SSH within (energy) research (e.g. Foulds and Christensen 2016; Sovacool 2014; Sovacool et al. 2015).

It is therefore clear that the SSH are being pursued directly as part of a particular configuration of interdisciplinarity that traverses both the Natural/Technical Sciences and the SSH. We believe that this pursuit is widely understood and observed by research stakeholders, although we strongly contend that the implications of this configuration (which funders and other actors alike are pushing) have not been given the attention it deserves. For instance, how may Natural/Technical Scientists imagine the role of SSH in their projects, and vice versa, and with what effects for a collaboration's power dynamics? How can knowledge be translated to become credible among distinct and hitherto separated cultures of scientific knowledge production? Indeed, the implications of such a marriage is a central thread of this book that we return to at various stages. Beyond this though, and aside from being a key part of our object of study (interdisciplinarity), we argue that the SSH themselves also have much to offer to the very study approaches utilised—including, for instance, positionality, infrastructures and epistemics of knowledge production, movement of knowledge, dynamics of appropriation, different disciplinary

[1] https://eassh.eu/.

[2] https://www.shapeid.eu/.

[3] This book's second author co-led the EU platform for energy-SSH via the EU Horizon 2020–funded projects: SHAPE ENERGY, over 2017–2019 (www.shapeenergy.eu); and Energy-SHIFTS, over 2019–2021 (www.energy-shifts.eu).

interpretations of scientific findings, and the importance of organising around disciplinary collectives. It is exactly in these respects that our next sub-section discusses our proposed role of Science and Technology Studies (STS) in better understanding the dynamics and experiences underlying research practice.

1.1.2   What Science and Technology Studies' Sociotechnical Underpinnings Have to Offer

In the previous sub-section, we made clear that calls and some structural support for interdisciplinary research exist within the mainstream management and delivery of research systems. We also discussed the role that SSH can play in shedding more light on the social dynamics of interdisciplinary research. In this sub-section, though, we take this further by specifically drilling into what the underutilised STS can offer the study of interdisciplinary practice. Not only is this the core rationale upon which this book is based, but this sub-section also implicitly represents this book's first call (of many) for those leading interdisciplinary project evaluations to directly engage with STS ideas. But first, it is necessary herein to step back and consider the origins of STS and, in particular, what STS itself offers through its common point of departure.

STS is a large and an increasingly popular and heterogeneous area of research, and while attempts to define the field for relative outsiders exist (e.g. Sovacool et al. 2020; STS Helsinki 2021), it is not in our interest to develop a concise designation here. Indeed, we contend that any such fixed designation would not serve the diverse corners of this area. In general terms, though, STS is unique in its interest in the interplay between science, technology, and society, in a manner that pays specific conceptual and empirical interest to the actual content and processes of science and technology in the making. This focus remains unique among social scientific studies, many of which still treat the social and technical aspects of science and technology as a binary and, in doing so, narrow the Social Sciences to the study of 'the social' dimensions of these matters. STS takes the binary distinction to be a fallacy, hence the common way to term its focus as being *sociotechnical* (Silvast et al. 2013). Therefore, to reiterate: being sociotechnical "is not the same as either just having social and technical researchers in the same research team or in having researchers trained in both disciplinary routes" (Cooper 2017, p. 115). Whilst these may indeed be happening within sociotechnical studies, actually being sociotechnical requires a fundamental, ontological appreciation of the *coproduction* of the social and the technical. For a more detailed review of the STS field, we recommend one of the recent handbooks, such as Felt et al. (2016).

Within these terms of reference, there is inevitably a range of STS interests and perspectives in play around the roles of science and technology in practice and normatively in society. In this book, though, we are particularly inspired by the Science Studies component of STS, which has for decades explored the social construction of (scientific) knowledge (e.g. Hacking 2000; Knorr Cetina 1999; Latour and Woolgar 1979; Pinch and Bijker 1984). STS itself has some of its origins in studies of professional scientists collaborating with one another in Natural Science laboratories, for example, and thus if we assume current societies to be ruled by expertise and knowledge, it is possible to start utilising such insights (previously used around, for instance, the knowledge society; Knorr Cetina 1999, Chapter 10) for investigating the knowledge-object relationships of interdisciplinary practice too.

Indeed, given the interest of STS in how "sets of relations" (Law 1991, p. 18) shape knowledge creation, for example, STS is no stranger to the study of interdisciplinarity. In fact, the reflexivity advocated for by STS scholars has been put to good use in considering how STS itself emerged as a discipline. For example, both Mitcham (2003) and Sørensen (2012) discussed the interdisciplinary 'disciplining' of STS to the extent that it became an interdiscipline, and Cozzens (2001) similarly argued that a unifying core of 'STS thought' could only ever exist once STS researchers (themselves usually from different disciplines) were able to leave their past disciplinary baggage behind them. Moreover, in preceding these discussions, there were even questions as to whether interdisciplinarity could be feasibly achieved within STS, given its disciplinary positioning and organisation (Bauer 1990). It is therefore evident that whilst STS explorations into interdisciplinarity remain in the minority, STS does have a track record of asking the deeper questions of interdisciplinarity.

To be clear, we assert that an STS perspective cannot be treated as a single entity, given that the interdiscipline is dispersed and itself positioned across and between disciplines. Here, through this book, we adopt the following considerations from it:

- That SSH studies of interdisciplinarity should focus on the actual content of scientific research in large-scale, collaborative projects.

- That this requires a set of research methods that capture this actual content, including conventional qualitative interviews and fieldwork, alongside any methods that dig into the mundane everyday dynamics of project planning and implementation.
- That while we detail the inner life of (interdisciplinary) projects, this content of science—including the ways in which projects are organised—has clear normative implications. Therefore, the study of a project's inner life does not only stop at its situated practices, but extends to the institutional terms and and contexts of actions that those practices sit within.

STS therefore offers tools to enable a deeper unpacking of interdisciplinarity in the making. In using STS to dig deeper into the more mundane everydayness of doing interdisciplinarity, a richer picture is generated that allows one to move beyond more simplistic discussions of identifying 'barriers', 'obstacles', 'challenges', and so on. It is these more simplistic discussions that can be overly reductive and linear, for instance, through implying that their identified barriers need only be jumped over or busted through to neatly 'fix' centuries-old institutions and ensure that interdisciplinary efforts will prosper (c.f. Shove 1998). Indeed, over the last 20 years, there has been a plethora of studies that have focused explicitly on identifying the barriers to doing interdisciplinarity (e.g. Brewer 1999; Campbell 2005; Cohen et al. 2021; Hein et al. 2018; Kelly et al. 2019; Lyall and Meagher 2012; Morse et al. 2007; Wallace and Clark 2017). Such studies are often situated within a wider *descriptive* convention that lacks a conceptual bedrock to their discussion of interdisciplinarity. We assert that this then commonly leads to the same sorts of generic difficulties and recommendations being reproduced relating to, for example, disciplinary languages, communication strategies, balancing expertise, resource burdens, and career trajectories. Whilst quite often rich and undoubtedly interesting in isolation, we strongly argue that such interdisciplinary studies are reaching their saturation in terms of their contributions. Our collective understanding of interdisciplinarity is not advancing at the rate that it once was. STS can help rectify this by filling the current conceptual void and by asking questions that have not yet been sufficiently explored.

Fundamentally: SSH has much to contribute to the study of interdisciplinarity, including (but not limited to) how SSH themselves are addressed within interdisciplinary approaches. STS specifically offers a solid, underutilised basis for moving beyond a mere descriptive account of the experiences encountered. Connecting said experiences within and through such conceptual tools allows scholars to speak more to, and learn from, other contributions in the literature, and, as such, better interpret interdisciplinary data at hand. In this vein, it is the drawing together of STS-led concepts that this book argues for and evidences the merits of—we now discuss this, in the context of this book's broader contributions and positionings, in the next section.

1.2   Introducing This Book

1.2.1   A Position Statement on Notions of Interdisciplinarity

To make clear exactly what we mean by 'interdisciplinarity' or 'interdisciplinary research' in this book, we now outline our positions on key boundaries and scope issues. This position statement outlines where we position ourselves amongst the diverse approaches to interdisciplinarity—including, for instance, what 'interdisciplinarity' even means and how it is operationalised as an object of study. Specifically, in this sub-section, we present five positions in turn, which together form the foundations of and offer context for many of this book's arguments.

1.2.1.1   Position #1: Definitions of Disciplines Should Account for the Interconnectedness, Porosity, and Inevitable Subjectivity of Their Knowledges and Knowledge-Making Communities

In considering our position on the boundaries and relations between disciplines, we believe it is important to first reflect on what a discipline is. Indeed, we contend that works on interdisciplinarity rarely contain any definition of a discipline—although perhaps this is to be expected, given that the rationale for these works is in transgressing disciplines. Interdisciplinary scholars may therefore be fundamentally critical of disciplines and hence potentially feel that they do not need to define what they critique. Whatever the reasoning, any critique or discussion of interdisciplinarity will suffer if one is not clear on one's terms of reference (e.g. scope, boundaries, and purpose) for a discipline.

In reflecting on what a discipline is, we found Jacobs' (2013, p. 28) discussion instructive: "A discipline is a form of social organization that generates new ideas and research findings, certifies this knowledge, and in turn teaches this subject matter". In drawing parallels between defining disciplines and defining professions, Jacobs developed this further through discussion of, for example, scholarly associations, conference participation, publishing strategies, career pathways, and responsibilities for handing over to the next generation, that together socially organise institutional disciplinary groupings.

What is also clear from Jacobs' (2013) discussion is the importance of acknowledging the messy interconnections between disciplines. Indeed, whilst disciplines can provide useful proxies for different ways of generating, interpreting, and applying knowledges, we should not obsess about them to the point where the porosity of disciplinary boundaries is forgotten. Knowledges and their associated institutional structures cannot be compartmentalised:

> [K]nowledge is transgressive. Nobody, in my awareness, has succeeded anywhere for very long in containing knowledge. It seeps through institutional structures like water through pores of a membrane. As with liquids in membranes, knowledge seeps in both directions. (Gibbons and Nowotny 2001, p. 68)

Rigidly drawing boundaries between academic disciplines is not always possible and/or useful. Discussion of interdisciplinarity would therefore only benefit from acknowledging that no objective categorisation will do justice to the disciplinary complexity and evolution in play. Indeed, we welcome the inclusion of disciplines—which are, themselves, constructions—that are self-assigned and self-identified. We do not believe it productive to rigidly apply top-down classifications of what a discipline can and should be, as it would close off possibilities of including new, emerging disciplines—which themselves may be hybrid disciplines (or 'interdisciplines') that may have arisen through a common set of interdisciplinary research interests (e.g. Gender Studies and Urban Studies).

Furthermore, there are many intersecting scales and dimensions as to how disciplines are organised. For example, is Environmental Social Science a discipline in itself, or does it just constitute part of Environmental Science? Contestation around disciplinary labels is inevitable, and this is wholly appropriate—it could never be possible to achieve consensus, not least because disciplines change, evolve, emerge, and fade, too. This certainly fits with the second author's experiences of, for example, producing disciplinary lists of researchers (e.g. SHAPE ENERGY 2017) and analysing open survey questions that ask for disciplinary associations (e.g. Foulds et al. 2017, p. 17).

An implication of acknowledging this interconnectedness between disciplines is that it directly problematises the assumption that the adding together of different, so-called 'distinct' disciplines will objectively add up to a 'complete' picture. Indeed, it is often assumed that interdisciplinarity represents the completion of a "jigsaw" (Castree and Waitt 2017, p. 3), where the connection of new disciplinary additions supposedly reveals more of "an 'objective world' awaiting discovery and accurate reporting" (Castree and Waitt 2017, p. 3). Thus, seeing disciplines in the way that we have set out above, then, has implications for our expectations of what interdisciplinary research can realistically achieve.

1.2.1.2   *Position #2: We Focus Primarily on Interdisciplinary Problem-Focused Research and Not on Interdisciplinary General Education*

To borrow from Klein's (1990) terms, the 'interdisciplinary general education' (p. 156) form of interdisciplinarity—which targets the pre-disciplinary mode of understanding—is not where we concern ourselves in this book. Instead, we primarily focus on what Klein (1990) refers to as 'interdisciplinary problem-focused research' (p. 121), specifically related to furthering societies' response to challenges associated with low-carbon sociotechnical transformations.

For this form of interdisciplinarity, and as discussed by Mitcham (2003), problem-focused interdisciplinarity has tended to originate via either (1) research-producing communities being interested in and subsequently posing new cognitive questions that span across disciplines or (2) an interest in generating (often technical) solutions for practical problems facing societies, which may or may not be pushed by the problem-holders themselves (e.g. policy actors).

As our Position #1 implied, the social organisation of disciplines cannot be neatly separated into different institutional activities. Yet, despite this, we believe it important to be clear on whether the ambitions underlying one's interdisciplinarity is more/less grounded in problem-focused interdisciplinary research or within educational approaches to broadening understanding (especially if said interdisciplinarity is acting as an object of research in itself). Our implicit focus on interdisciplinary problem-focused research therefore makes clear that our discussion of interdisciplinarity in this book inevitably contains certain normative dimensions.

1.2.1.3 Position #3: The Full Spectrum of Multidisciplinary, Interdisciplinarity and Transdisciplinarity Should Be Part of a Broad Definition of Interdisciplinarity That Covers the Range of Crossdisciplinary Research Practice

It has long been said that interdisciplinarity lacks a coherent, single definition (Salter and Hearn 1996). As Callard and Fitzgerald (2015, p. 4) put it, "interdisciplinarity is a term that everyone invokes and none understands". We certainly note that multidisciplinarity and transdisciplinarity are also used frequently, and are often conflated and/or used interchangeably with interdisciplinarity. In distinguishing between each of these terms, we note the following (inspired by Klein [2010] in particular):

- *Multidisciplinary research:* parallel endeavours from different disciplines, which do not have (or at least do not prioritise) integration.
- *Interdisciplinary research:* integrated perspectives from different disciplines that add up to more than the sum of their parts.
- *Transdisciplinary research:* a deeper degree of integration than interdisciplinarity, to the point where different disciplines are more deeply 'fused', leading to clear opposition and/or a new alternative to established disciplinary conventions. These new conventions may often involve the pursuit of normative goals, based around real-world problems (Lawrence and Després 2004). It is in this way that the starting point for transdisciplinarity is sometimes talked about as not being dependent on pre-existing disciplines, unlike interdisciplinarity which does firmly start from those pre-existing disciplinary standpoints and considers how best integration can be organised between them.

Some use the additional term of 'crossdisciplinarity', but we argue that all of interdisciplinarity, multidisciplinarity, and transdisciplinarity are forms of crossdisciplinary research practice. Whilst literature around these terms have been useful in certain respects, as part of carving out deeper reflection as to one's positionality on the wide spectrum of crossdisciplinary research practice, we firmly agree with Petts et al. (2008), who note that "at its weakest, interdisciplinarity constitutes barely more than cooperation, while at its strongest, it lays the foundation for a more transformative recasting of disciplines" (Petts et al. 2008, p. 597). We therefore side with scholars such as Barry et al. (2008, p. 28), who "take 'interdisciplinarity' as a generic term for this spectrum, while signalling salient issues from the definitional debate as they arise". A lot of what we discuss throughout this book on interdisciplinarity is therefore relevant too for the debates on transdisciplinarity and—although perhaps to a lesser extent due to the lower levels of integration—multidisciplinarity. All of these crossdisciplinary endeavours share common ideals and aspirations: to bring monodisciplinary communities together in novel ways to generate fruitful and integrated insights.

We thus continue to use the term 'interdisciplinarity' through this book as a catch-all term for crossdisciplinary research practice. When we do use the terms 'transdisciplinarity' or 'multidisciplinarity', we do so intentionally as part of emphasising a particular point—usually in contrast to what would be the case for 'typical' interdisciplinarity.

1.2.1.4 Position #4: Interdisciplinarity Does Not Only Occur in the Space Between More Technical/Natural and More Social Scientific Disciplines

Putting aside debates on the spectrum of disciplinary integration that may occur, at its most basic level it is important to note that 'interdisciplinarity' (in our catch-all sense) is simply about bringing two or more disciplines together. As such, no disciplines have exclusive rights on participating in interdisciplinarity. Interdisciplinarity can occur, for instance, within the Social Sciences and Humanities (SSH), between, say, Sociology and History, or it could bridge across an SSH discipline and a Technical/Natural Science discipline, say, between Human Geography and Civil Engineering—either is just as valid.

We make this point to ensure lines of enquiry remain open to the dynamics that are co-produced by bringing together the various combinations of both 'near' and 'far' disciplines. Each discipline will imagine another discipline in particular ways, and thereby come to expect certain outcomes, and indeed it is based on those expectations that different configurations of interdisciplinarity will feel more or less comfortable to prospective participants.

This said, we do acknowledge that interdisciplinarity is predominantly regarded by funders and policymakers as being a bridge between far disciplines; in this case, between the more technical/natural scientific and the more social scientific approaches. For example, the EU has focused on 'mainstreaming' SSH disciplines across all of its Horizon 2020–funded research on 'societal challenges' (Kania and Bucksch 2020), supporting its underlying view that interdisciplinarity is a means to overcome the non-technical (SSH) barriers for scientific solutions to prosper (c.f. Guy and Shove 2000). It is therefore in prioritising the implications of these agendas—and in unpicking the ground-level experiences of their implementation—as to why the sorts of interdisciplinarity covered in this book predominantly relate to the integration of the (energy-related) Technical/ Natural Sciences on the one hand and the (energy-related) SSH on the other hand.

1.2.1.5 Position #5: Interdisciplinarity Can Include, But Does Not Necessitate, the Involvement of External Stakeholders

Discussions with colleagues have regularly involved suggestions that multi-stakeholder engagement represented interdisciplinarity. Seemingly, such arguments were based on the assumption that working across sectors was the same as working across disciplines. We would strongly argue that this is not the case; a sector should not be conflated as being equal to a discipline, regardless of any parallels that can be drawn between professional and disciplinary jurisdictions.

Nevertheless, we do note that certain disciplinary configurations of interdisciplinarity are more open to multi-stakeholder involvement than others, especially when compared to many monodisciplinary approaches. We also note that different disciplinary configurations and forms of multi-stakeholder engagement will bring with them different norms and conventions for working with stakeholders (and this is reflected in our examples later in this book). Such considerations matter in making the point that interdisciplinarity can welcome, but does certainly not necessarily require, stakeholder engagement.

It is certainly true that the crossdisciplinary approaches typically termed as being transdisciplinary would require the active participation of different stakeholders (Winskel 2018), but we argue that integrating stakeholders into one's plans is not in itself interdisciplinary.

1.2.2   Headline Contributions: Aim and Scope of This Book

The aim of this book is to develop an STS framework for examining interdisciplinarity in the making. In fulfilling this aim, we make four contributions, which we now briefly discuss in turn.

First and foremost to our core aim, we provide a *Sociology of Interdisciplinarity*, where we detail a new framework that is of use both to those new to interdisciplinarity (in all its various configurations and guises) and to those who have been working interdisciplinarily for many years. Fundamentally, we put the spotlight on overlooked issues that have not yet entered mainstream discourse on interdisciplinarity—whether in, among others, researcher, funder, or policy communities. Our framework is succinctly based around six key dimensions. There is much to be gained by stepping back to sociologically consider the collective commonalities of (interdisciplinary) research practice.

Second, this *Sociology of Interdisciplinarity* is primarily inspired by the work of STS literatures. We use STS to unpack interdisciplinary research in practice and explain its successes and failures sociologically. We strongly contend that STS has considerable potential for developing one's understanding of interdisciplinarity, not least because it has a proven track record of studying the co-evolutions of professions and professional practice within the messy entanglements of the social and the material (e.g. Knorr Cetina 1999; Latour and Woolgar 1979). Such studies range, for example, across various professional domains: scientific laboratories (e.g. Latour and Woolgar 1979), domestication of new technologies (e.g. Lie and Sørensen 1996), and the biographies of artefacts (e.g. Hyysalo 2021), to name only a few. Using STS as the foundations to our framework also ensures a consistent, and obviously appropriate, sociotechnical ontology for further dialogue on interdisciplinarity.

Indeed, whilst there has been some discussion of the normative role that interdisciplinary research plays as part of a "logic of ontology" (Barry et al. 2008, p. 25) that aims to drive ontological change in/across existing disciplines, there has been very little (public) discussion on the ontological logics that underlie the research focused on interdisciplinary practice itself. This lack of explicit ontological consideration and/or foregrounding is symptomatic of interdisciplinary studies focusing too much on, for example, debating taxonomies, describing barriers, producing generic recommendations, or considering interdisciplinarity only as a social problem. We argue that the conceptual underpinnings behind studies of interdisciplinarity should be placed within the broader SSH debates on the fundamentals of what makes up social order and governs social action. Adopting a consistent ontological line, with support from its associated conceptual tools, will ultimately allow for a richer discussion on interdisciplinarity, and it is in this regard that we present STS as an underutilised option.

Third, we apply our *Sociology of Interdisciplinarity* framework to matters of (interdisciplinary) energy research. Indeed, during the last decade, the ideal of interdisciplinary research has enjoyed strong support in energy research (Winskel 2018) and among European (Kania and Bucksch 2020) and several national funding agencies (e.g. Norwegian Research Council 2018; UK Engineering and Physical Council Sciences Research 2021). This book provides a detailed STS-inspired examination of how interdisciplinary energy research has been conceived, and with what consequences and dynamics for those involved in such projects. Furthermore, as per our previous assertions for STS and interdisciplinarity more generally, we similarly contend that STS has been markedly underutilised in the study of interdisciplinarity in energy research. Indeed, STS has been used to frame questions for specific (interdisciplinary) matters on energy system transformations (e.g. Hess and Sovacool 2020; Hyysalo 2021; Hyysalo et al. 2018; Jalas et al. 2017), but it has not yet been used to investigate the interdisciplinary practice underlying the pursuit of researching those energy system transformations.

Fourth, in exploring the aforementioned issues, we will draw on rich empirics. Specifically, through Chaps. 2, 3, and 4, we bring fresh insights into the lived experiences and actual contents of large-scale, collaborative energy research projects. Through this, we delve into interdisciplinarity directly or at least consider some of the interdisciplinary struggles associated with monodisciplinarity, and thus we do not restrict ourselves to merely advocating for interdisciplinarity and/or our particular *Sociology of Interdisciplinarity* framework. We believe we have interesting stories to tell that can help bring our conceptual discussions to life—and this is of particular use to those readers who may be firmly interested in interdisciplinary and/or energy research, but with less of a background in STS.

Finally, in taking inspiration from MacKenzie's (2009) introductory remarks to his own STS-focused framework, we similarly argue that our book builds up to a set of dimensions that are implicitly agreeable to STS and related critical-SSH communities, even if those dimensions are not yet widely used in the study of interdisciplinarity. Such potential agreement is perhaps inevitable given how our arguments are fundamentally linked to STS' shared point of departure. Nevertheless, again like MacKenzie (2009, p. 4), we appreciate that the approach we construct and advocate through this book is inevitably "idiosyncratic" and "'incomplete", and we would therefore not wish to "foist" our ideas onto our colleagues. Instead, we hope that our contributions represent the start of further work in this area; this book is not intended to close down debate and discussion (as, e.g. a definitive end-point), but rather to prompt critique, extension, and further empirical consideration from others.

1.2.3   Structure and Journey of This Book

In the context of our own interpretation of and positioning on what inter-disciplinarity exactly is (Sect. 1.2.1) and in delivering our stated contributions (Sect. 1.2.2), the remainder of this book proceeds as follows: Chaps. 2, 3, and 4 represent the empirical core of this book, within which we discuss the first author's experiences in three large-scale energy research projects. Specifically, Chap. 2 discusses the dynamics of working interdisciplinarily within UK whole systems research on energy and brings to the fore what energy modellers expect from SSH scholars. Chapter 3 reflects upon the evolution and organisation of Norwegian environment-friendly energy research centres and in doing so particularly emphasises the importance of funding structures in funnelling certain configurations of interdisciplinarity. Chapter 4 then intentionally offers a different empirical perspective—a more conventional, monodisciplinary reference point, from which this book's core interests in interdisciplinarity can be contextualised—in a bid to further progress our argument on route to this book's conclusions. It considers a large Finnish, monodisciplinary research project on the pricing of energy risks, and provides complementary insights on issues of objectivities, power dynamics, science-policy translations, and interdisciplinarity roadblocks. All these empirical insights from Chaps. 2, 3, and 4 directly feed into our proposition for a *Sociology of Interdisciplinarity* (Chap. 5), where we present six dimensions: the impacts of funding; epistemic cultures; boundary objects; appropriating disciplines; interpretative flexibility; and the importance of disciplines.

Ultimately, this book applies critical social scientific ideas to the study of interdisciplinarity, relating in particular to the use, deployment, and appropriation of SSH disciplines within large-scale energy research projects. More specifically, we utilise approaches to interdisciplinarity that are directly inspired by STS. We are therefore especially interested in the practices and materiality of interdisciplinarity, including, for example, the importance of objects, technologies, and equipment (e.g. computer models), as well as the embeddedness of human actors in this materiality. Indeed, actors are constructed in certain ways as part of developing and maintaining interdisciplinary collaborations.

REFERENCES

Barry, A., Born, G., 2013. Interdisciplinarity, in: Barry, A., Born, G. (Eds.), Interdisciplinarity: Reconfigurations of the social and natural sciences. Routledge, London, pp. 1–56.

Barry, A., Born, G., Weszkalnys, G., 2008. Logics of interdisciplinarity. Economy and Society 37, 20–49. https://doi.org/10.1080/03085140701760841

Bauer, H., 1990. Barriers against interdisciplinarity: Implications for studies of science, technology, and society (STS). Science, Technology, & Human Values 15, 105–119.

Brewer, G.D., 1999. The challenges of interdisciplinarity. Policy Sciences 32, 327–337.

British Academy, 2016. Crossing paths: Interdisciplinary institutions, careers, education and applications. British Academy, London.

Callard, F., Fitzgerald, D., 2015. Rethinking interdisciplinarity across the social sciences and neurosciences. Palgrave Macmillan, London.

Campbell, L.M., 2005. Overcoming obstacles to interdisciplinary research. Conservation Biology 19, 574–577. https://doi.org/10.1111/j.1523-1739.2005.00058.x

Caniglia, G., Luederitz, C., von Wirth, T., Fazey, I., Martín-López, B., Hondrila, K., König, A., von Wehrden, H., Schäpke, N.A., Laubichler, M.D., Lang, D.J., 2021. A pluralistic and integrated approach to action-oriented knowledge for sustainability. Nature Sustainability 4, 93–100. https://doi.org/10.1038/s41893-020-00616-z

Castree, N., Waitt, G., 2017. What kind of socio-technical research for what sort of influence on energy policy? Energy Research and Social Science 26, 87–90. https://doi.org/10.1016/j.erss.2017.01.023

Cohen, J.J., Azarova, V., Klöckner, C.A., Kollmann, A., Löfström, E., Pellegrini-Masini, G., Gareth Polhill, J., Reichl, J., Salt, D., 2021. Tackling the challenge of interdisciplinary energy research: A research toolkit. Energy Research & Social Science 74, 101966. https://doi.org/10.1016/j.erss.2021.101966

Cooper, A.C.G., 2017. Building a socio-technical energy research community: Theory, practice and impact. Energy Research & Social Science 26, 115–120. https://doi.org/10.1016/j.erss.2017.02.001

Cozzens, S.E., 2001. Making disciplines disappear in STS, in: Cutcliffe, S.H., Mitcham, C. (Eds.), Visions of STS: Counterpoints in science, technology, and society studies. SUNY Press, Albany, NY, pp. 51–64.

European Commission, 2019. How should interdisciplinarity and stakeholder knowledge be addressed and evaluated in Horizon 2020 proposals? [WWW Document]. URL https://ec.europa.eu/info/funding-tenders/opportunities/portal/screen/support/faq/935 (accessed 16.8.21).

European University Association, 2017. Energy transition and the future of energy research, innovation and education: An action agenda for European Universities. EUA, Brussels.

Felt, U., Fouché, R., Millar, C.A., Smith-Doerr, L. (Eds.), 2016. The handbook of science and technology studies, Fourth. ed. MIT Press, Cambridge, MA.

Foulds, C., Christensen, T.H., 2016. Funding pathways to a low-carbon transition. Nature Energy 1. https://doi.org/10.1038/nenergy.2016.87

Foulds, C., Robison, R., 2018. Mobilising the energy-related social sciences and humanities, in: Foulds, C., Robison, R. (Eds.), Advancing energy policy: Lessons on the integration of social sciences and humanities. Palgrave Macmillan, Cham, pp. 1–12.

Foulds, C., Robison, R., Balint, L., Sonetti, G., 2017. Headline reflections— SHAPE ENERGY call for evidence. SHAPE ENERGY, Cambridge.

Foulds, C., Royston, S., Berker, T., Nakopoulou, E., Abram, S., Ančić, B., Arapostathis, E., Badescu, G., Bull, R., Cohen, J., Dunlop, T., Dunphy, N., Dupont, C., Fischer, C., Gram-Hanssen, K., Grandclément, C., Heiskanen, E., Labanca, N., Jeliazkova, M., Jörgens, H., Keller, M., Kern, F., Lombardi, P., Mourik, R., Ornetzeder, M., Pearson, P., Rohracher, H., Sahakian, M., Sari, R., Standal, K., Živčič, L., 2020. 100 social sciences and humanities priority research questions for energy efficiency in Horizon Europe. Energy-SHIFTS, Cambridge.

Gibbons, M., Nowotny, H., 2001. The potential of transdisciplinarity, in: Thompson Klein, J., Grossenbacher-Mansuy, W., Häberli, R., Bill, A., Scholz, R.W., Welti, M. (Eds.), Transdisciplinarity: Joint problem solving among science, technology, and society. Birkhäuser Basel, Basel, pp. 67–80. https://doi.org/10.1007/978-3-0348-8419-8_7

Guy, S., Shove, E., 2000. A sociology of energy, buildings, and the environment: Constructing knowledge, designing practice. Routledge, London.

Hacking, I., 2000. The social construction of what? Harvard University Press, Cambridge, MA.

Hein, C.J., Ten Hoeve, J.E., Gopalakrishnan, S., Livneh, B., Adams, H.D., Marino, E.K., Susan Weiler, C., 2018. Overcoming early career barriers to interdisciplinary climate change research. Wiley Interdisciplinary Reviews: Climate Change 9, e530. https://doi.org/10.1002/wcc.530

Hess, D.J., Sovacool, B.K., 2020. Sociotechnical matters: Reviewing and integrating science and technology studies with energy social science. Energy Research & Social Science 65, 101462. https://doi.org/10.1016/j.erss.2020.101462

HM Government, 2017. Industrial strategy: Building a Britain fit for the future. UK Government Department for Business, Energy & Industrial Strategy, London.

Hulme, M., 2018. "Gaps" in climate change knowledge. Environmental Humanities 10, 330–337. https://doi.org/10.1215/22011919-4385599

Hyysalo, S., 2021. Citizen Activities in Energy Transition: User Innovation, New Communities, and the Shaping of a Sustainable Future, First. ed. Routledge, Abingdon.

Hyysalo, S., Juntunen, J.K., Martiskainen, M., 2018. Energy internet forums as acceleration phase transition intermediaries. Research Policy 47, 872–885. https://doi.org/10.1016/j.respol.2018.02.012

Irwin, E.G., Culligan, P.J., Fischer-Kowalski, M., Law, K.L., Murtugudde, R., Pfirman, S., 2018. Bridging barriers to advance global sustainability. Nature Sustainability 1, 324–326. https://doi.org/10.1038/s41893-018-0085-1

Jacobs, J.A., 2013. In Defense of Disciplines: Interdisciplinarity and Specialization in the Research University. University of Chicago Press, Chicago.

Jalas, M., Hyysalo, S., Heiskanen, E., Lovio, R., Nissinen, A., Mattinen, M., Rinkinen, J., Juntunen, J.K., Tainio, P., Nissilä, H., 2017. Everyday experimentation in energy transition: A practice-theoretical view. Journal of Cleaner Production 169, 77–84. https://doi.org/10.1016/j.jclepro.2017.03.034

Jasanoff, S., 2013. Fields and Fallows: A political history of STS, in: Barry, A., Born, G. (Eds.), Interdisciplinarity: Reconfigurations of the social and natural sciences. Routledge, London, pp. 99–118.

Kania, K., Bucksch, R., 2020. Integration of social sciences and humanities in Horizon 2020: Participants, budgets and disciplines—5th monitoring report on projects funded in 2018 under the Horizon 2020 programme. European Commission Directorate-General for Research and Innovation, Brussels.

Kelly, R., Mackay, M., Nash, K.L., Cvitanovic, C., Allison, E.H., Armitage, D., Bonn, A., Cooke, S.J., Frusher, S., Fulton, E.A., Halpern, B.S., Lopes, P.F.M., Milner-Gulland, E.J., Peck, M.A., Pecl, G.T., Stephenson, R.L., Werner, F., 2019. Ten tips for developing interdisciplinary socio-ecological researchers. Socio-Ecological Practice Research 1, 149–161. https://doi.org/10.1007/s42532-019-00018-2

Klein, J.T., 2010. A taxonomy of interdisciplinarity, in: Thompson Klein, J., Mitcham, C., Frodeman, R. (Eds.), The oxford handbook of interdisciplinarity. Oxford University Press, Oxford, pp. 15–30.

Klein, J.T., 1990. Interdisciplinarity: History, theory, and practice. Wayne State University Press, Detroit, MI.

Knorr Cetina, K., 1999. Epistemic cultures: How the sciences make knowledge. Harvard University Press, Cambridge.

Latour, B., Woolgar, S., 1979. Laboratory life: The construction of scientific facts. Sage Publications, Beverly Hills.

Law, J., 1991. Introduction: Monsters, machines and sociotechnical relations, in: Law, J. (Ed.), A sociology of monsters: Essays on power, Technology and Domination. Routledge, Abingdon, pp. 1–23.

Lawrence, R.J., Després, C., 2004. Futures of transdisciplinarity. Futures 36, 397–405. https://doi.org/10.1016/j.futures.2003.10.005

Lie, M., Sørensen, K.H. (Eds.), 1996. Making technology our own? Domesticating technology into everyday life. Tanum, Oslo.

Lyall, C., Meagher, L.R., 2012. A Masterclass in interdisciplinarity: Research into practice in training the next generation of interdisciplinary researchers. Futures 44, 608–617. https://doi.org/10.1016/j.futures.2012.03.011

MacKenzie, D., 2009. Material markets: How economic agents are constructed. Oxford University Press, Oxford.

Mitcham, C., 2003. Toward an STS experiment with interdisciplinarity. Bulletin of Science, Technology & Society 23, 473–478. https://doi.org/10.1177/0270467603261348

Morse, W.C., Nielsen-Pincus, M., Force, J., Wulfhorst, J., 2007. Bridges and barriers to developing and conducting interdisciplinary graduate-student team research. Ecology and Society 12, 8.

Nature, 2015. Why interdisciplinary research matters. Nature 525, 305. https://doi.org/10.1038/525305a

Norwegian Research Council, 2018. FME—Forskningssentre for miljøvennlig energi [WWW Document]. URL https://www.forskningsradet.no/om-forskningsradet/programmer/fme/ (accessed 12.31.19).

Pedersen, D.B., 2016. Integrating social sciences and humanities in interdisciplinary research. Palgrave Communications 2, 16036. https://doi.org/10.1057/palcomms.2016.36

Pellerin-Carlin, T., Vinois, J.-A., Rubio, E., Fernandes, S., 2018. Making the energy transition a European success: Tackling the democratic, innovation, financing and social challenges of the Energy Union. Jacques Delors Institute, Brussels.

Petts, J., Owens, S., Bulkeley, H., 2008. Crossing boundaries: Interdisciplinarity in the context of urban environments. Geoforum 39, 593–601. https://doi.org/10.1016/j.geoforum.2006.02.008

Pinch, T.J., Bijker, W.E., 1984. The social construction of facts and artefacts: Or how the sociology of science and the sociology of technology might benefit each other. Social Studies of Science 14, 399–441. https://doi.org/10.1177/030631284014003004

Robison, R., Foulds, C., 2021. Social sciences and humanities for the European Green Deal. 10 recommendations from the EU Energy SSH Innovation Forum. Energy-SHIFTS, Cambridge.

Robison, R., Foulds, C., 2019. 7 principles for energy-SSH in Horizon Europe: SHAPE ENERGY Research & Innovation Agenda 2020–2030. SHAPE ENERGY, Cambridge.

Royston, S., Foulds, C., 2021. The making of energy evidence: How exclusions of Social Sciences and Humanities are reproduced (and what researchers can do about it). Energy Research & Social Science 77, 102084. https://doi.org/10.1016/j.erss.2021.102084

Salter, L., Hearn, A., 1996. Outside the lines: Issues in interdisciplinary research. McGill-Queen's University Press, Montreal and Kingston.
Science Europe, 2019. Symposium report: Interdisciplinarity. Science Europe, Brussels.
SHAPE ENERGY, 2017. Researcher Database [WWW Document]. URL https://shapeenergy.eu/index.php/researcher-database/ (accessed 7.19.17).
Shove, E., 1998. Gaps, barriers and conceptual chasms: theories of technology transfer and energy in buildings. Energy Policy 26, 1105–1112. https://doi.org/10.1016/S0301-4215(98)00065-2
Silvast, A., Hänninen, H., Hyysalo, S., 2013. Energy in society: Energy systems and infrastructures in society. Science & Technology Studies 26, 3–13. https://doi.org/10.23987/sts.55285
Sørensen, K.H., 2012. Disciplined interdisciplinarity? A brief account of STS in Norway. TECNOSCIENZA 3, 49–61.
Sovacool, B.K., 2014. What are we doing here? Analyzing fifteen years of energy scholarship and proposing a social science research agenda. Energy Research & Social Science 1, 1–29. https://doi.org/10.1016/j.erss.2014.02.003
Sovacool, B.K., Hess, D.J., Amir, S., Geels, F.W., Hirsh, R., Rodriguez Medina, L., Miller, C., Alvial Palavicino, C., Phadke, R., Ryghaug, M., Schot, J., Silvast, A., Stephens, J., Stirling, A., Turnheim, B., van der Vleuten, E., van Lente, H., Yearley, S., 2020. Sociotechnical agendas: Reviewing future directions for energy and climate research. Energy Research & Social Science 70, 101617. https://doi.org/10.1016/j.erss.2020.101617
Sovacool, B.K., Ryan, S.E., Stern, P.C., Janda, K., Rochlin, G., Spreng, D., Pasqualetti, M.J., Wilhite, H., Lutzenhiser, L., 2015. Integrating social science in energy research. Energy Research & Social Science 6, 95–99. https://doi.org/10.1016/j.erss.2014.12.005
STS Helsinki, 2021. What is STS? [WWW Document]. URL https://blogs.helsinki.fi/sts-helsinki/what-is-sts/ (accessed 5.11.21).
UK Engineering and Physical Council Sciences Research, 2021. Whole energy systems [WWW Document]. URL https://epsrc.ukri.org/research/ourportfolio/researchareas/wholesystems/ (accessed 4.29.21).
UKRI, 2021. Interdisciplinary research [WWW Document]. URL https://re.ukri.org/research/interdisciplinary-research/ (accessed 4.29.21).
University of Essex, 2020. Interdisciplinary Studies Centre: Student Handbook 2020–2021. University of Essex, Colchester.
Wallace, R.L., Clark, S.G., 2017. Barriers to interdisciplinarity in environmental studies: A case of alarming trends in faculty and programmatic wellbeing. Issues in Interdisciplinary Studies 35, 221–247.
Winskel, M., 2018. The pursuit of interdisciplinary whole systems energy research: Insights from the UK Energy Research Centre. Energy Research & Social Science 37, 74–84. https://doi.org/10.1016/j.erss.2017.09.012

CHAPTER 2

Whole Systems Thinking and Modelling in the UK

Abstract UK academic researchers have been vying for a 'whole' systems perspective on energy issues for more than a decade. This research programme has exposed challenges in complex systems thinking and in the dialogue between academic disciplines and epistemic cultures that is needed to mediate the social, technological, and environmental impacts of energy systems. This chapter examines these efforts starting from existing studies that include detailed reports on experiences of interdisciplinary research. By extending these findings via interviews and ethnographic research, this chapter pays particular attention to the role of interdisciplinary computer modelling that was expected to represent complex energy transitions and energy infrastructures of the future. In doing so, this chapter demonstrates how interdisciplinarity has actually worked in three exemplary areas: the diversity of computer models that seek to represent everyday energy demand and how they simplify both demand and other disciplines in so doing; the need for collaborative, cross-cutting research in foresight of future energy scenarios; and how modelling scholars strongly envision their models should become 'useful' for imagined policy and planning stakeholders.

Keywords National Centre for Energy Systems Integration (CESI) • UK Research and Innovation (UKRI) • Computer modelling • Modelling evidence • Policy • Ethnography • Social Sciences and Humanities

2.1   Introduction

UK academic researchers, assigned to the task by national research funder priorities, have been vying for a 'whole' systems perspective on energy issues for more than a decade. This holistic, and somewhat aspirational, concept of systems has several relevant roots. In particular, the academic field of Systems Engineering has been instrumental in its development, not least because it fundamentally comprises the study and management of whole systems, through the application of interdisciplinary knowledge. It relies on systems theory that emerged in the twentieth century, although it never became a single unified theory of complexity (see Labanca et al. 2020 for further details). Such contributions in recent decades have also evidenced the link between complex systems approaches and the building of energy systems, such as electricity networks and other large infrastructures (Hughes 1983; van der Vleuten 2004).

The concept of whole energy systems is an umbrella term in the UK, evidently developed actively among research funders (e.g. the umbrella organisation UK Research and Innovation, UKRI—previously Research Councils UK, RCUK), research projects (e.g. Transition Pathways consortium and National Centre for Energy Systems Integration, reviewed below), and networks of academics (e.g. UK Energy Research Centre, UKERC). From this starting point, one early designation of whole systems outlines energy research that involves "thinking about all the dimensions of change and drawing on a range of disciplines and expertise" (UKERC 2009, p. 5), where the dimensions are society, economy, and the environment, and the aim is understanding the complex challenges of such holistic energy systems. It situated an interdisciplinary focus that corresponded with a wide definition of an energy system as "the set of technologies, physical infrastructure, institutions, policies and practices located in, and associated with the UK which enable energy services to be delivered to UK consumers" (UKERC 2009, p. 16).

The UK whole systems research programme—of which UKERC's original work was instrumental in summarising—has subsequently exposed new possibilities and challenges in complex systems thinking, as well as stimulated dialogue between academic disciplines and epistemic positions. This chapter is exactly interested in examining these efforts. Indeed, the core empirical focus of this chapter is on the experiences of (mainly Social Scientists) working in a large-scale predominantly technical research project, the National Centre for Energy Systems Integration (CESI). CESI seeks to develop more integrated energy systems in the UK through the design of more integrated and interactive 'whole' energy systems for the future. Funded by the UK Engineering and Physical Sciences Research Council (EPSRC), the project involves five research universities and several industrial partners (running 2016–2022). We conducted fieldwork within CESI and in university research groups more generally to access a wider group of energy researchers.[1]

The fieldwork was conducted over two university terms between 2017 and 2018. It included 12 interviews with academic modellers, from mainly two of CESI's UK research universities. By discipline, almost all the interviewees were based in Engineering and the Physical Sciences, although we note that some started their research careers outside of the energy field, in a variety of disciplines, from Architecture and Astrophysics to Applied Mathematics. Three interviewees (25%) were female, and the rest were male. Nearly all participants were either Postdoctoral or PhD Researchers, and therefore relatively junior members of staff in universities, which we acknowledge is a limitation of the data: more senior staff would have had different perspectives on designing and leading interdisciplinary projects, and early-career researchers are at a known risk from doing interdisciplinary research (Lyall 2019). This empirical shortcoming is mitigated by drawing upon not only the interviews, but also insights from participation in various project events (especially workshops) and from grey project papers—that is to say, we sought insights from situations and materials where senior members of staff may have voiced concerns about interdisciplinarity.

Using original fieldwork and existing reports, this chapter therefore specifically aims to unpack the dynamics of whole systems projects and how they have worked, according to the scholars participating in them. In discussing the experiences of participants in heterogeneous UK research projects, this chapter pays particular attention to the role of interdisciplinary computer modelling (with a specific focus on quantitative planning, operational, and demand tools, as well as qualitative narrative scenarios) that was expected to represent complex energy systems integration, including energy infrastructures of the future.

[1] The first author thanks Prof. Simone Abram, the Principal Investigator of this study at Durham University, for leading the work and conducting some of this field research together. Partly the same materials have been used in our earlier publications (Silvast et al. 2020), but this chapter is a further elaboration in line with the argument of this book focused on interdisciplinary knowledge production.

This chapter is structured as follows: we begin by contextualising our discussion by drawing key lessons from previous interdisciplinarity-focused evaluations of large UK energy research projects (Sect. 2.2). We then analyse the working practices of CESI in three selected research examples, all of which highlight the need for interdisciplinary research and integration across the Social Sciences and Engineering. The first example is energy demand modelling, the second future scenarios, and the third policy relevance and planning (Sect. 2.3). This chapter concludes by beginning to gather the elements of the framework of this book, which is deepened in the subsequent empirical chapters (Sect. 2.4), on route to its final presentation and discussion in Chap. 5.

2.2   Reviewing Past Experiences: Interdisciplinary Working Practices in Whole Systems Energy Research

The UK research projects addressed here have a dual relationship to interdisciplinarity: they have been facilitating interdisciplinary research, but they also produced evaluations of how these disciplinary integrations have worked. These insider reports come from a variety of project contexts and seniorities, indicating the breadth and the diversity of interdisciplinary energy research activities in the UK. While one report examines the experiences of early-career researchers across several UK energy demand research projects, with a particular focus on building research (Mallaband et al. 2017), others were produced as part of particular energy project, such as the first phase of the Transition Pathways to a Low Carbon Economy project[2] funded by the EPSRC (2008–2012) that interviewed a wide range of academic seniorities (Hargreaves and Burgess 2010; Longhurst and Chilvers 2012). Meanwhile, another report (Winskel et al. 2015) examined the whole of UKERC's interdisciplinary capacities and achievements between 2004 and 2014. Across these reports, we contend that there are four common messages that emerge—we now discuss each of these in turn.

---

[2] The Transitions Pathways project was awarded another phase in 2012 to continue under the banner "Realising Transition Pathways—Whole Systems Analysis for a UK More Electric Low Carbon Energy Future" until 2016, a few months before the CESI project started. All the reports used here, in this chapter, were produced in the first phase of the project.

Firstly, interdisciplinarity as a concept and practice found wide support not only among research funders, but also among project participants, whether their broad discipline was in Engineering or the Social Sciences and Humanities (SSH). That said, more salient differences existed with regard to what the participants anticipated from said interdisciplinarity. For example, positions varied from simply gathering interdisciplinary teams in order to broaden perspectives, to more deeply facilitating integrated problem framings—these two positions were linked to different expectations of Engineers and SSH scholars respectively (Hargreaves and Burgess 2010; Longhurst and Chilvers 2012). Consequently, the depth of interdisciplinary collaboration varied in these projects, depending on research output and site, and how exactly to do this disciplinary integration in practice was not always agreed upon (Mallaband et al. 2017).

Secondly, and related to this first message, interdisciplinarity requires an active effort to make a project work coherently for its participants, which is in some tension to how interdisciplinarity should also be about open exploration that should not be made to cohere with *ex ante* agendas too strongly (Winskel et al. 2015). Moreover, a project's coherence is not a single entity and should not be treated as a simplistic aim. Instead, Longhurst and Chilvers (2012) highlighted four different meanings of coherence in interdisciplinary energy research: the coherent linkage of different knowledge outputs; the comprehensions between project participants; coherence as the credibility of interdisciplinary research outputs; and the more conceptual coherence between different disciplinary worldviews. These forms of coherence also find their corollary in the CESI project examined in the Sect. 2.3.

Thirdly, the early-career researchers in particular (Mallaband et al. 2017), but partly also all the academics interviewed for these reports, highlighted the influence of expectations and roles in the everyday lives of projects. Namely, there are different expectations both concerning what interdisciplinarity is meant to achieve and which roles SSH scholars in particular take and/or are afforded in these projects. This finding joins the insight that has emerged from other fields, such as Synthetic Biology, where one paper (Balmer et al. 2015) recognises no less than nine different roles that SSH scholars can assume in interdisciplinary collaborations. Only one of these roles is a conventional academic research colleague, which highlights that interdisciplinary experts can remain at a greater or smaller distance from the core (still disciplinary-based) activity of the said projects.[3]

Fourthly, and importantly for the argument of this book, interdisciplinary working has institutional interdependencies that are recognised by project participants. This means that institutional and research funder support is needed for successful interdisciplinary research—an issue that is not merely about individual achievements among the project members (Mallaband et al. 2017). Part of this problem, as is now well known in interdisciplinarity literature (e.g. Lyall 2019), is that interdisciplinarity is highly favoured as a label but not always served by discipline-based academic reward systems, in terms of, for example, careers, funding, and publications. Indeed, UKERC (Winskel et al. 2015) reiterate the key role that research funding, commissioning, and assessment take in supporting research capacity in interdisciplinarity. They specifically call for collective responsibility of these actors and partnerships among academics, and for them to further pursue these achievements.

In summary, we see that interdisciplinarity requires active maintenance and the (re)production of coherence. We also can observe that the SSH do not always have a preconfigured role in these collaborations, but that this role is instead actively constructed by funding calls and the project participants themselves. The depth and the coherence of interdisciplinary working also vary depending on the kinds of research output or situation at hand.

2.3   A New Whole Systems Approach and Energy Integration Issues

In furthering our discussion, this section now turns to in-depth findings from a relevant large-scale interdisciplinary project (CESI), to reflect on how such knowledge production efforts happen in practice, and with what implications for the pursuit of a whole systems' understanding. We begin by detailing the project background and modelling contexts that the CESI project operates within (Sect. 2.3.1), before then discussing three main findings: the diversity of interdisciplinary models and the case of energy demand (Sect. 2.3.2); the challenges of prediction and the consequences

---

[3] The other roles are 'the representative of the public', 'the foreteller', 'the wife', 'the critic', 'the trickster', 'the reflexivity inducer', 'the educator', and 'the co-producer of knowledge'.

of doing interdisciplinarity (Sect. 2.3.3); and interdisciplinary planning under infrastructural conditions (Sect. 2.3.4).

2.3.1   Modelling Within the National Centre for Energy Systems Integration Project

The starting point of the CESI project concerned computer modelling in energy. In this, the project was not entirely novel: modelling methods and interdisciplinary discussions on them had been central to various whole systems projects, including both phases of the Transition Pathways to a Low Carbon Economy (Hargreaves and Burgess 2010) and the UKERC Energy 2050 project (UKERC 2009). Computer models are formal representations of bounded energy systems that combine Mathematics and Computing. These tools model envisioned energy systems and can be used to anticipate the energy systems of the future and the present (Silvast et al. 2020).

One well-known model in the UK, although not in use in CESI, is the MARKAL[4] model (Taylor et al. 2014). The MARKAL model was initiated by the International Energy Agency countries in the 1970s. Within the UK, it was actively maintained and developed by the UCL Energy Systems Team in University College London (UCL 2021). The MARKAL model has been widely used both in the UK and internationally, among governments as well as academics, and for a long period of time until the late 2010s. Its successor model, TIMES,[5] is very similar. This is how UCL characterises the MARKAL model:

> MARKAL portrays the entire energy system from imports and domestic production resources (fossil and renewable), through fuel processing and supply (e.g., refining, bio-processes), explicit representation of infrastructures (e.g., gas pipelines), conversion of fuels to secondary energy carriers (including electricity, heat and hydrogen), end-use technologies (residential, commercial, industry, transport, agricultures, non-energy), and energy service demands (at a sub-sectoral level) for the entire UK energy system. (UCL 2021, n.p.)

By presenting this highly complex energy system by a model, the MARKAL seeks to optimise the system: in practical terms, for example, to

---

[4] MARKAL: MARKet and ALlocation.
[5] TIMES: The Integrated MARKAL-EFOM System.

minimise energy systems costs while constrained by given physical and policy dimensions of energy. The CESI project—which, as already noted, did not develop or use MARKAL—emerged around similar modelling efforts and promised to progress a wide interdisciplinary programme around them. Initially, CESI research noted how current modelling approaches were insufficient, namely because:

> [they] are unable to provide sufficiently accurate or detailed, integrated representations of the physics, engineering, social, spatial temporal or stochastic aspects of real energy systems. They also struggle to generate robust long-term plans in the face of uncertainties in commercial and technological developments and the effects of climate change, behavioural dynamics, and technological interdependencies. (UKERC Energy Data Centre 2021, n.p.)

This description links together social and technical aspects of energy and includes a clear need for the Social Sciences (and, more implicitly, Humanities) researchers to engage in envisioning this future integrated system. Therefore, in integrating Engineering and the Physical Sciences to carry out modelling, CESI also ended up employing several SSH scholars, including Economists, Anthropologists, Geographers, and policy-facing energy researchers. An SSH-led stream of enquiry was hence designed into CESI's structure.

While the energy integration terms come from a related background of Systems Engineering (O'Malley et al. 2016), the link between this new energy integration project and whole systems thinking has also been evident. For example, Professor Phil Taylor, the founder and former director of CESI, has noticed a growing "consensus that 'a whole systems approach' is necessary to transform the UK energy system and drive forward the government's industrial strategy" (quoted in Northern Gas Networks 2017).

Sections 2.3.2, 2.3.3, and 2.3.4 recount how scholars working towards bringing about these 'whole systems'—in particular inside CESI, but also in related project environments—developed tools and methods for integrating academic disciplines and saw the possibilities and limits of their tools in bringing this integration about. Particular attention is given to computer models as means by which the interdisciplinary whole systems are being enacted.

2.3.2 *The Diversity of Interdisciplinary Models and the Case of Energy Demand*

The first key finding from the ethnography of whole systems modelling concerned just how *different* computer models are, and that there are various subgroups of modelling scholars that bring these differences about. In a general sense, this observation is not novel. The literature on energy research has provided numerous classifications of 'modelling families' in energy, which comprise different approaches, academic fields, goals, and differences among bottom-up and top-down models (Li et al. 2015; Pfenninger et al. 2014). However, what is novel here is bringing this diversity of models to the question of interdisciplinary knowledge production: what does the well-recognised diversity among energy models imply for integrating academic disciplines in knowledge production with models?

To enable our following discussion to be more tangible, and less abstract, we now use a seemingly straightforward topic as our object of enquiry: energy demand. In this modelling setting, demand is usually considered as final energy consumption. Nevertheless, as another large UK interdisciplinary energy consortium, CREDS,[6] has described it, energy demand is a wide topic. Demand is regarded as comprising all human activity that depends on available energy, shaping energy infrastructure as a result. Accordingly, demand "drives the whole energy system, influencing the total amount of energy used; the location of, and types of fuel used in the energy supply system; and the characteristics of the end use technologies that consume energy" (CREDS 2021 no pagination).

This much is clear to the current social scientific works on demand and everyday practices (e.g. Shove and Trentmann 2018). But how can energy modellers represent this demand with their quantitative tools? A recent UKERC working paper (Hardt et al. 2019) examined the energy models that inform UK government energy policy and enquired how these mainstream models represent energy demand. They generated a range of findings, with 13 core models selected covering different approaches, purposes, and parts of the energy system. While the model types also differed—ranging from econometric to system-optimisation, economic, and sector-based models—they clearly shared an emphasis on providing large technological detail as the main output. However, all the reviewed models also showed "limitations with regard to the representation of non-technological drivers

---

[6] CREDS: Centre for Research into Energy Demand Solutions.

of energy demand" (Hardt et al. 2019, p. 6). It was especially apparent that drivers of demand (e.g. behavioural, economic, and social dynamics) were both treated as exogenous assumptions in the models and did not belong to the objective of the modelling. This is interesting insofar as the studied modellers saw that their discipline could not offer any proof on demand, but this also made the studied models inadequate for understanding how demand is constituted and could change.

The report (Hardt et al. 2019) did not study academic modelling per se—as they focused only on how academic models are used in UK policy processes—but they did subject considerable expectation to interdisciplinary academic work that could fill the knowledge gap on demand. The authors anticipate that:

> there is still considerable scope for energy models to provide better representations of demand-side energy policies, especially with regard to non-technological aspects. The academic literature contains some promising attempts of incorporating more realistic representations of social and behavioural processes in energy models. (Hardt et al. 2019, p. 6)

This quote closely resembles the problem that the first author was asked to address in June 2018, with a few weeks to work on it before a project workshop: namely, to explore how models in Energy Systems Engineering could become more sociotechnical, by including qualitative and quantitative information on demand for energy. This was not an unknown problem to CESI. It involved a whole research group of 'demand modellers', that is, modellers dealing with everyday energy demands, and who had in many ways crossed over to the SSH. We saw these researchers actively exploring and using sociological practice theories in their talks, for instance, and seeking to complement their models with our 'Social Science data' on everyday life to make the results more uncertain or what was termed as 'fuzzy' and on occasions, even labelling their work as explicitly social scientific, or as the looser common term in use goes, 'sociotechnical' (cf. Love and Cooper 2015).

For example, the first author attended a project meeting where one of the CESI demand modellers named his presentation as "socio-technical energy demand modelling". The presenter's research question asked how people and institutions are represented in energy systems models, and the presentation itself travelled from qualitative narrative scenarios to modelling of behaviour, the rise of active energy citizens, and how social practices confront abrupt changes in societies. While the speaker remained ambivalent on his discipline being a Social Science, in the discussion that ensued, the contribution of the presentation was explicitly addressed as social scientific by some external participants of the workshop.

However, while CESI did have a team of demand modellers, this did not mean that there was a single way to accomplish this integration of demand to modelling. As the demand modellers themselves clearly understood, modelling demand had to confront considerable uncertainties surrounding everyday practices. Indeed, as one CESI modeller observed:

> [In a] Traditional model you put discrete information about the control of the system, something comes on at certain time, goes off at a certain time, you can play about with temperature settings and so on, but there is discrete information … The fuzziness of what happens, lots of people using those things where they might start this time, or they might start this time, or this time … It's a bit difficult to put in manually into this model. [Man, 40–49 years, Senior Lecturer in Energy)

More than saying that some models could not appropriate uncertain demand—which was true for some, but not all of them—it was more accurate that demand, in itself, had 'interpretative flexibility' (Pinch and Bijker 1984). Different scientists offered not merely different interpretations of what demand may be, but also different designs for how demand could be included in their models. An explorative classification, based on the fieldwork, found these families of demand approaches in CESI:

1. *The demand curve approach*: demand is treated as external to the modelled energy system, but interacts with it as its 'environment' or 'input'. This demand can be represented, for instance, by actual long-term measurement on energy use. This is close to the models identified (Hardt et al. 2019) that also use demand as exogenous to the model.
2. *The known demand approach*: demand is already 'known' by the model. It can be represented by actual energy use data or simulated by an algorithm. This 'known' constant of the model is fixed, where other variables (e.g. voltage and temperatures) are solved by the model. The difference to the demand curve approach is that, here, demand is not exogenous, but is internal to the modelled system.

3. *The techno-economic approach*: demand is represented by assuming consumers are rational economic actors that, for example, respond to price signals. This assumption exists behind many future planning models and considerations on introducing dynamic (e.g. time-of-use) electricity pricing to households. It is important to note that many models assume these rational consumers to actually exist, even if this assumption has been widely challenged in the SSH for decades (cf. Christensen et al. 2020).

4. *The demand modelling approach*: demand becomes the actual output of the modelling work. This could happen by measuring it empirically in households or simulating it, or often via a combination of both. This approach persists, for instance, in many building modelling studies where modellers seek to know how energy demand evolves in certain kinds of buildings and with a set of improvements to, for example, energy efficiency.

5. *The impacts to demand approach*: the model results, whether the models include demand or not, will always enact certain kinds of energy technologies and energy systems, often in a simulated representation. This approach to demand means enquiring what those changed technologies and systems would imply for the everyday demand of people and the activities and habits that constitute it.

6. *The demand foresight approach*: especially when it comes to anticipating future energy systems, it is important to know how society's demands for energy may change in the coming decades, for example, up to 2050 in line with governmental decarbonisation targets. This approach to demand is common in scenarios and storylines that merge 'qualitative' with 'quantitative' modelling, which will be discussed more in Sect. 2.3.3.

While overlapping and related to one another, these approaches to the concept of demand are not the same, and there may be no middle ground that would integrate all of them. To a varying extent, most of them clearly call upon integration from the SSH to produce insights on people's activities that constitute demand. But depending on which approach to demand is taken, the concepts, methodologies, and evidence bases from SSH would be almost entirely different. The provision of more accurate demand curves might call expertise from, for example, the Statistics and Economics disciplines of SSH. Whereas enquiries on how new energy systems will affect people's lives might need insights and tools from field methods in

Anthropology and Sociology; or Law and Policy Studies when the legal and regulatory dimensions of these changes are of interest. Further, the questions relating to how our societies will develop in 2050 could be addressed by forecasting methods. Such different contributions would ultimately mean entirely different tasks for SSH engagement in modelling projects.

The above findings also suggest that modelling—while largely based across the disciplines of Engineering, Economics, Statistics, and Computer Science, and in itself is firmly interdisciplinary—cannot be unified as a single (inter)discipline in its own right. The notion of a discipline remains important for this interdisciplinary energy work, especially when scholars attribute what belongs to their expertise and what does not, which is typical when considering demand and human activities. But disciplines do not entirely capture the different ways in which modellers produce knowledge, again, for example, about demand. Resembling 'epistemic cultures' (Knorr Cetina 1999), different modellers had various tools and methods, types of reasoning, ways to establish evidence, and ideals concerning theory-empirics fits. These differences extended to how their models could integrate demand, including reflections on to what extent it was even possible. Those engaging in interdisciplinary collaborations with modellers should pay attention to these differences as much as the disciplinary differences, for example, between core Engineering and SSH, which are communicated far more often in our experience.

Yet another conclusion to be made from the findings above is just how distinctly the modellers and policy studies of models attributed the role of SSH. If we assume that demand is the domain of the SSH—and this was routinely assumed by the modelling experts and we have never heard it questioned by them—then it is appropriate to ask, which SSH scholars would recognise that they study, for example, 'non-technological drivers of demand' or 'social and behavioural processes'? The answer is that these are labels attributed to the SSH from the outside. However, the issue runs deeper than the process of labelling, because these labels are symptoms of wider institutional norms and deeper-running expectations. As such, SSH scholars cannot (or, rather, should not) aim to simply replace uncomfortable labels with more appropriately deemed alternatives.

Here, we can draw on work positioned between Anthropology and Information Science, to discuss a more significant knowledge gap (Forsythe 1999). The idea that patterns of behaviour and organisation can be integrated to computer models also assumes that those patterns are out there, just waiting to be detected by SSH-led observation. Yet, the issue is that many current SSH methodologies do not provide answers to such a problem. It is now well known that topics (e.g. behaviour and organisation) are not just 'out there' in the social world, but constructed during the research process, and our methods not only describe them but also bring them about, often during a meticulous and long research process (Gobo 2008; Silvast and Virtanen 2019). This makes it challenging to work in interdisciplinary modelling projects partly in hidden ways. If the research problems have been designed in such a way where an appropriate SSH response is difficult, this might require engaging the partners on what SSH exactly is and is not, which is very rarely (if ever) among the deliverables of common research and innovation projects.

2.3.3   The Challenges of Prediction and the Consequences of Doing Interdisciplinarity

A second key finding, where the role of interdisciplinary working becomes manifested, concerns foreseeing the energy systems of the future. This is the core domain of large energy research projects, with the European Union and several national governments having set decarbonisation goals up to 2050, and researchers seeking sociotechnical solutions to help reach those goals. Many computer models, although not all of them, are committed to foreseeing how this future energy system will come about. Earlier relevant UK projects, such as the Transition Pathways (Longhurst and Chilvers 2012), were explicitly committed to scoping this complex sociotechnical change in an interdisciplinary manner. This was also true of CESI, whose main research aim was to understand both future energy supply and demand.

The notion of predicting energy futures has also become increasingly problematised, which modellers themselves know much about. Here, the underlying issue is uncertainty and including it in the modelling process in a useful way. Indeed, past academic and policy predictions have had demonstrable issues in dealing with uncertainty, and are thus starting points for considering improvements. For instance, a UKERC retrospective study of UK energy forecasts discovered past models to have been poor at incorporating future uncertainties, and many real-world events that had followed these predictions would have been considered to be too extreme when making the predictions (McDowall et al. 2014).

Most of the studied future scenarios had only emphasised economic issues (e.g. predicting oil prices) and focused on specific technologies (e.g. nuclear power). They were focused on cost-optimisation and none of the examined scenarios would allow for significant institutional changes (McDowall et al. 2014). Indeed, institutional arrangements, political decisions, and the impacts of societal actors may be difficult to include in a useful modelling process, although some influential attempts exist nonetheless (Li et al. 2015). A problem related to this is that energy models are not modelling a fixed target system. Instead, the energy systems being modelled—including their regulations, technologies, business models, and end-use practices—are undergoing change simultaneously (McDowall 2014). This situation challenges the prospects of predicting how those systems will behave in the future.

The general response to these issues of forecasting has been to reformulate the aim: towards aspiring for 'good-quality', rather than 'accurate', predictions of the future. We have come far from the scenario-building and forecasting exercises of the 1970s and the 1980s, where calculative models would be used to make deterministic energy supply policies and were even performative to the notion of what energy policy is all about (Aykut 2019). Today's academic energy futures are rather named as storylines—that is, narrative stories and visions (Fortes et al. 2015)—and there is an explicit assumption that storylines are not predictions or forecasts, and should not be treated as such.

Typically, storylines designate that certain events will happen in the future, such as "macroeconomic and microeconomic policies simultaneously stimulate innovation, creativity, and technological improvement" (Fortes et al. 2015, p. 164). Yet, the scenario method does not actually predict that this will happen; in some cases and depending on the method, it does not claim it is more likely, nor does it have to claim that it is more favourable than any another envisioned scenario, such as the innovation not being stimulated. The starting point is rather in developing storylines towards challenging decision-makers, and these stories can also be linked with information from quantitative models in a myriad of ways.

CESI was also similarly exploring qualitative and quantitative scenarios with the aim of improving strategic insights for energy systems integration. To this aim, it set about developing narrative scenarios for the UK. At the time of writing this, the scenarios are a work-in-progress and it is not on our agenda to comment more on effective scenario design or different approaches to scenarios, but we do draw on a grey paper including several

CESI members (Wheatcroft et al. 2019). We will simply note that the CESI approach to scenarios seems to resemble what was outlined above, especially the impossibility of perfect prediction of future events and scepticism when it comes to applying probabilities to future scenarios. Rather than predicting, the paper by CESI and others outlines five aims for a scenario (Wheatcroft et al. 2019, p. 6):

1. *Plausible:* a scenario should be plausible and come with a narrative justifying each event or change in the underlying assumptions.
2. *Distinctive:* the different scenarios should be distinctive enough in terms of the key factors for there to be a clear difference between them.
3. *Consistent:* interaction between key factors should be taken into account. For example, macroeconomic factors may impact important aspects of the scenarios simultaneously.
4. *Relevant:* each scenario should be relevant in terms of giving a specific insight into the future (e.g. the government increases spending on green projects and subsidies).
5. *Challenging:* scenarios should challenge the conventional view on things that may affect the project in question.

This listing moves us back to our interest in interdisciplinary working practices. The authors of the CESI paper are particularly interested in mainstream mathematical and statistical modelling and their relationships with scenarios. However, what role will SSH play in formulating the scenarios of the future?

Here, we notice a very different pathway depending on whether scenarios are meant to be predictions or not. If scenarios are just like predictions, then SSH research can be deployed to discover whether the predictions seem to be realistic or not, and why. Economists and Social Scientists could work together to find out how likely, for example, it is that macroeconomic policies work in stimulating innovations in different cases, and how that may be changing. However, the problem, as the writers of this aforementioned CESI paper (Wheatcroft et al. 2019) see it, is that scenarios and storylines are not meant to be predictive, and there is a delicate (and not adequately communicated) difference between predicting and simply offering plausible explanations. But, what would this difference mean in operational terms? Here, more work is needed to explain the terms in a way that clarifies the difference to various involved disciplines.

We also see a risk that the SSH become an add-on that merely challenges the scenarios developed by other disciplines (Robison and Foulds 2019), but this does not yet constitute strong interdisciplinary integration. We would claim that to take the five recommendations above to their logical conclusion—to form scenarios that are plausible, distinctive, consistent, relevant, and challenging—requires SSH scholars to be integrated into the work of forming scenarios from its start, which was always the case in CESI. Enquiring what is, for example, plausible about energy futures is a complex task and requires different expertise across disciplines, including various kinds of SSH scholars, from Anthropology to Philosophy, Ethics, Political Science, and beyond (cf. Ialenti 2020). Here, we again voice the need for going beyond simple prescriptions, such as SSH scholars only study the 'non-technological drivers of demand'. Adequate and balanced disciplinary representation (Winskel et al. 2015) is needed to ensure that energy scenarios' definitions of the future are not shifted towards biases developed by other dominant disciplines.

2.3.4   Interdisciplinary Planning Under Infrastructural Conditions

Our third finding was that more than 'accuracy' or 'prediction', the CESI modellers were visibly more interested in how their models are used and by whom. This became especially pronounced in the contexts of policy, politics, and planning—contexts set out with the borders of the computer models, yet meant to be informed by modelling results (Silvast et al. 2020). In contrast to the two other areas mentioned in the previous two sub-sections—where the SSH were often implicitly present, by topics that were assumed to be SSH-relevant, but were not always mentioned by name—in this case, the role of SSH was more explicit and pronounced.

Herein, we will use a workshop report, published by Centre for Digital Built Britain scoping network, called *Planning Complex Infrastructure Under Uncertainty*, and including several CESI members, as an exemplar (Dent et al. 2019).[7] At its outset, the report summarises the work of "researchers in mathematical sciences, engineering and social sciences" (Dent et al. 2019, p. 1). In addition, further than this, the report's one

---

[7] For disclosure, the first author was one of the workshop participants, took part in the discussions, and is named in the report.

core recommendation was to involve both SSH in the planning of infra-structures. The report states this aim:

> There is a need to incorporate social science (including humanities) research around issues such as understanding 'value', capturing change in value/s, multiplicity of voices (success for whom?), and critical assessments of data and models. There are different potential relationships available between social science research and that of science, mathematics and engineering, and scope to consider these relationships creatively in developing interdisci-plinary work; there is value in social science research not only to support and/or challenge work in technical subjects but also sometimes to lead or shape the challenges addressed and approaches taken. (Dent et al. 2019, p. 4)

There is much packed into this quote. It starts by designating certain implicitly 'non-technical' tasks for SSH, such as increasing the diversity of values and voices in planning infrastructures, and critiquing the models and data prepared presumably by non-SSH disciplines. But the end of the quote shows considerably more variation to the relationships between dis-ciplines. It even envisions that, in some cases, the disciplinary balance could be turned the other way, to let SSH scholars shape research projects on infrastructural matters.

Another topic where the SSH are more implicitly present comes to the policy relevance of modelling tools. While focused around the topics of modelling techniques and research, the report is also written in a way that it clearly seeks to translate the modelling practice into new areas. A par-ticular reliance is placed on models that support decision-making and that are even designed with decision-support in mind. As the authors express this aim:

> It is important to guard against matters such as collecting data for the sake of having a large dataset, or confusing optimality in the model world with a good decision in the real world—the real goal being to identify decisions which one has logical reason to believe are good ones in the real world. (Dent et al. 2019, p. 9)

In our ethnography, we observed a widely shared similar interest: in decision-making, decision-support, and the 'appropriate' use of models by these decision-makers. Curiously, however, this role of decision-support is not where the SSH were recognised above. The report still states a divi-sion where the SSH predominantly study the 'non-technical', and perhaps also the 'non-political' aspects of energy, or study the political insofar as it is translated into values and opinions. Not mentioning decades of SSH insights on governance, political studies, and policy analysis, the model designers appropriated the needs of the model 'end-users' on their common sense: using assumptions about how 'policy decisions' are made, rather than in-depth knowledge of governance practices. As we have suggested elsewhere (Silvast et al. 2020), training in governance would be an appropriate step for modellers working in energy-policy-interfaces and seeking to engage policymakers. A further integration of models and policy would also require the existence of models that can act as 'boundary objects' (Star and Griesemer 1989)—such as the popular MARKAL model in the UK, which is understood across different social worlds (e.g. academia and energy policy) and has hence become widely deployed (Taylor et al. 2014). All of this shows that the integration of the social worlds of policy, SSH, Engineering, and so on is not a simple task and requires more than explicit proclamations of interdisciplinary work to happen successfully.

2.4   Conclusions

This first empirical chapter of the book studied UK whole systems energy modellers and scientists: scholars who had explicitly set out to integrate natural, environmental, social, and technical disciplines, in coming to more relevant solutions to current energy issues.

The first main section (Sect. 2.2) of this chapter recounted how the issues of this research have been known, and documented, for a decade, often by reports from these projects from within. The next main section (Sect. 2.3) sought to study aligned themes in one of the most recent whole systems research programmes in the UK: the National Centre for Energy Systems Integration (CESI). CESI had designed SSH into its enquiry, but relied strongly on energy computer modelling in conjunction with interdisciplinary SSH research. Based on an ethnography conducted within the CESI and relying on the fieldwork and grey papers published by the CESI members, we travelled through three findings. The first finding was the diversity of modelling approaches and resulting variety of SSH integrations within these, using the case of energy demand modelling across the CESI and in UK policy more generally (Sect. 2.3.2). The second finding looked at how energy futures are being envisioned by interdisciplinary energy projects and what role SSH scholars could play in constructing a balanced view of these futures (Sect. 2.3.3). The final finding examined how CESI itself had vied for including SSH in planning complex infrastructures under uncertainty, and what that had implied for the possibilities of other disciplines to work, for example, on the policy relevance of the modelling results (Sect. 2.3.4).

We now summarise a set of six lessons learned from this first empirical chapter, which we hope further signals our direction of travel, on route to our end-of-book synthesis where we outline our *Sociology of Interdisciplinarity* framework. Our first lesson: UK research funding has had significant effects in bringing about more interdisciplinary agendas and certain kinds of working practices and collaborations when it comes to interdisciplinary working (e.g. Winskel et al. 2015). While we cannot study the structuring impact of funding with the data that we have here— for example, for scientific productivity (Goldfarb 2008)—we do argue that the research of individual scholars and groups is organising around themes that are of high priority in addressing grand societal challenges when it comes to energy (Royston and Foulds 2021). This chapter has started to document how individuals and groups working on these interdisciplinary projects develop joint work addressing the themes in their own ways and contexts. This is a theme we will continue to develop throughout the book.

Second, it is useful to think about interdisciplinary knowledge production in specific epistemic cultures (Knorr Cetina 1999), by which we mean knowledge-oriented cultures of scholars that cut across broad academic disciplines, such as Engineering, Physics, and the SSH. This view helps point out that disciplines—while highly relevant for reasons we point below—are not quite complex enough as units of analysis for understanding, for example, the diversity of knowledge production tools that (energy) researchers use. An example is the considerable variety that exists within energy modelling.

Third, the epistemic cultures in interdisciplinary projects are mediated by specialised boundary objects (Star and Griesemer 1989), such as computer models, conceptions of energy demand, or energy scenarios, as studied in this chapter. Boundary objects are artefacts, concepts, or methods that lie at the interface of different social worlds, such as politics and the academia. Because their identity is understood across these worlds—even in cases where they lack a proper definition—they enable co-operation and coordination between them. For instance, this comes very close to how the modellers examined in this chapter hoped that their models would work: as an object by which SSH scholars, Engineers, Mathematicians, policymakers, and various other actors alike could interact across their social worlds, even without always understanding the intricacies of how the models work.

Fourth, we see not only co-operation between different disciplines, but dynamics of appropriation. Interdisciplinary projects can see one discipline appropriating the tools and methods of other disciplines. This appropriation does not have to imply power dynamics or one discipline being more powerful than the other, and it does not have to be embedded in resource distribution (although it can be). Instead, in our observations, appropriation happened in much more mundane ways, especially relating to the labelling of what SSH do in modelling projects, without consulting what their research designs can actually allow. This generates the view that SSH only study, for example, 'social and behavioural processes', thereby implying activities that few SSH scholars would perhaps recognise and which would require much more demanding research resources than may be given to them for project implementation.

Fifth, while interdisciplinarity is often favoured by funding bodies and researchers as a label, this conceals the considerable interpretative flexibility of the concept itself. This premise extends to how those working in interdisciplinary projects interpret certain important interdisciplinary concepts (e.g. scenarios, energy demand, and energy policymaking). This finding is an extension of boundary objects, but makes a different conclusion: when scholars interpret concepts differently (e.g. energy demand), it means not only that the meaning differs, but that they would design technologies (e.g. demand modelling) in a distinct manner. This has major implications for how academic disciplines can work together for studying such concepts.

Sixth, our final lesson from this chapter: even in contexts where interdisciplinary is highly valued as an explicit strategy of the funding body, we should pay close attention to the continued importance of conventional academic disciplines in interdisciplinary work. There are more general reasons for the continued importance of disciplines, such as traditions of research offering coherence and presupposed practices, and students being taught in their paradigmatic instruments (Michael 2017). But the matter is also manifest in mundane project life and outputs: our materials show scholars routinely referring to broad academic disciplines as one homogenous entity and thereby with reference to a singular name (e.g. Mathematics, Statistics, Engineering, and Social Sciences), as part of them making sense of their own work in relation to others (Foulds et al. 2017).

We must take this continued use of disciplines by name seriously, since they clearly still mean much even for those engaged in projects that are meant to crossdisciplinary boundaries.

References

Aykut, S.C., 2019. Reassembling energy policy: Models, forecasts, and policy change in Germany and France. Science and Technology Studies 32, 13–35. https://doi.org/10.23987/sts.65324

Balmer, A.S., Calvert, J., Marris, C., Molyneux-Hodgson, S., Frow, E., Kearnes, M., Bulpin, K., Schyfter, P., MacKenzie, A., Martin, P., 2015. Taking roles in interdisciplinary collaborations: Reflections on working in post-ELSI spaces in the UK synthetic biology community. Science and Technology Studies 28, 3-25. https://doi.org/10.23987/sts.55340

Christensen, T.H., Friis, F., Bettin, S., Throndsen, W., Ornetzeder, M., Skjølsvold, T.M., Ryghaug, M., 2020. The role of competences, engagement, and devices in configuring the impact of prices in energy demand response: Findings from three smart energy pilots with households. Energy Policy 137, 111142. https://doi.org/10.1016/j.enpol.2019.111142

CREDS, 2021. What is energy demand [WWW Document]. URL https://www.creds.ac.uk/what-is-energy-demand/ (accessed 5.30.21).

Dent, C., Anyszewski, A., Reynolds, T., Masterton, G., Du, H., Tehrani, E., Lovell, K., Mackerron, G., 2019. Planning complex infrastructure under uncertainty—Network final report. https://doi.org/10.17863/CAM.40455.

Forsythe, D.E., 1999. "It's just a matter of common sense": Ethnography as invisible work. Computer Supported Cooperative Work 8, 127–145. https://doi.org/10.1023/A:1008692231284

Fortes, P., Alvarenga, A., Seixas, J., Rodrigues, S., 2015. Long-term energy scenarios: Bridging the gap between socio-economic storylines and energy modeling. Technological Forecasting and Social Change 91, 161–178. https://doi.org/10.1016/j.techfore.2014.02.006

Foulds, C., Robison, R., Balint, L., Sonetti, G., 2017. Headline reflections—SHAPE ENERGY Call for Evidence. Cambridge.

Gobo, G., 2008. Doing ethnography. Sage, London.

Goldfarb, B., 2008. The effect of government contracting on academic research: Does the source of funding affect scientific output? Research Policy 37, 41-58. https://doi.org/10.1016/j.respol.2007.07.011

Hardt, L., Brockway, P., Taylor, P., Barrett, J., Gross, R., Heptonstall, P., 2019. Modelling demand-side energy policies for climate change mitigation in the UK: A rapid evidence assessment. UKERC, London.

Hargreaves, T., Burgess, J., 2010. Pathways to interdisciplinarity: A technical report exploring collaborative interdisciplinary working in the Transition Pathways consortium, Working Paper—Centre for Social and Economic Research on the Global Environment. University of East Anglia, Norwich.

Hughes, T.P., 1983. Networks of power: Electrification in Western Society, 1880–1930. Johns Hopkins University Press, Baltimore.

Ialenti, V., 2020. Deep time reckoning: How future thinking can help Earth now. MIT Press, Cambridge, MA.

Knorr Cetina, K., 1999. Epistemic cultures. Harvard University Press, Cambridge, MA. https://doi.org/10.2307/j.ctvxw3q7f

Labanca, N., Pereira, Â.G., Watson, M., Krieger, K., Padovan, D., Watts, L., Moezzi, M., Wallenborn, G., Wright, R., Laes, E., Fath, B.D., Ruzzenenti, F., de Moor, T., Bauwens, T., Mehta, L., 2020. Transforming innovation for decarbonisation? Insights from combining complex systems and social practice perspectives. Energy Research and Social Science 65. https://doi.org/10.1016/j.erss.2020.101452

Li, F.G.N., Trutnevyte, E., Strachan, N., 2015. A review of socio-technical energy transition (STET) models. Technological Forecasting and Social Change 100, 290–305. https://doi.org/10.1016/j.techfore.2015.07.017

Longhurst, N., Chilvers, J., 2012. Interdisciplinarity in transition? A technical report on the interdisciplinarity of the Transitions to a Low Carbon economy consortium. University of East Anglia, Norwich.

Love, J., Cooper, A.C.G., 2015. From social and technical to socio-technical: Designing integrated research on domestic energy use. Indoor and Built Environment 24. https://doi.org/10.1177/1420326X15601722

Lyall, C., 2019. Being an interdisciplinary academic, being an interdisciplinary academic. Palgrave Macmillan, London. https://doi.org/10.1007/978-3-030-18659-3

Mallaband, B., Wood, G., Buchanan, K., Staddon, S., Mogles, N.M., Gabe-Thomas, E., 2017. The reality of cross-disciplinary energy research in the United Kingdom: A social science perspective. Energy Research and Social Science 25, 9-18. https://doi.org/10.1016/j.erss.2016.11.001

McDowall, W., 2014. Exploring possible transition pathways for hydrogen energy: A hybrid approach using socio-technical scenarios and energy system modelling. Futures 63, 1–14. https://doi.org/10.1016/j.futures.2014.07.004

McDowall, W., Trutnevyte, E., Tomei, J., Keppo, I., 2014. UKERC energy systems theme reflecting on scenarios. UKERC, London.

Michael, Mike., 2017. Actor-network theory: Trials, trails and translations. Sage, London.

Northern Gas Networks, 2017. Northern gas networks and CESI launch unique gas and whole systems research laboratory—IntEGReL [WWW Document]. URL https://www.northerngasnetworks.co.uk/2017/01/26/northern-gas-networks-and-cesi-launch-unique-gas-and-whole-systems-research-laboratory-integrel/ (accessed 5.30.21).

O'Malley, M., Kroposki, B., Hannegan, B., Madsen, H., Andersson, M., William, D., Mcgranaghan, M.F., Kroposki, B., Hannegan, B., Madsen, H., Andersson, M., Dent, C., 2016. Energy systems integration: Defining and describing the value proposition, Nrel/Tp-5D00-66616. https://doi.org/10.2172/1257674

Pfenninger, S., Hawkes, A., Keirstead, J., 2014. Energy systems modeling for twenty-first century energy challenges. Renewable and Sustainable Energy Reviews 33, 74–86. https://doi.org/10.1016/j.rser.2014.02.003

Pinch, T.J., Bijker, W.E., 1984. The social construction of facts and artefacts: Or how the sociology of science and the sociology of technology might benefit each other. Social Studies of Science 14, 399–441. https://doi.org/10.1177/030631284014003004

Robison, R., Foulds, C., 2019. 7 principles for Energy-SSH in Horizon Europe: SHAPE ENERGY Research & Innovation Agenda 2020–2030. Cambridge.

Royston, S., Foulds, C., 2021. The making of energy evidence: How exclusions of Social Sciences and Humanities are reproduced (and what researchers can do about it). Energy Research and Social Science. https://doi.org/10.1016/j.erss.2021.102084

Shove, E., Trentmann, F., 2018. Infrastructures in practice, infrastructures in practice. Routledge, London. https://doi.org/10.4324/9781351106177

Silvast, A., Laes, E., Abram, S., Bombaerts, G., 2020. What do energy modellers know? An ethnography of epistemic values and knowledge models. Energy Research and Social Science 66, 101495. https://doi.org/10.1016/j.erss.2020.101495

Silvast, A., Virtanen, M.J., 2019. An assemblage of framings and tamings: Multi-sited analysis of infrastructures as a methodology. Journal of Cultural Economy 12, 461–477. https://doi.org/10.1080/17530350.2019.1646156

Star, S.L., Griesemer, J.R., 1989. Institutional ecology, 'translations' and boundary objects: Amateurs and professionals in Berkeley's Museum of Vertebrate Zoology, 1907–39. Social Studies of Science 19, 387–420. https://doi.org/10.1177/030631289019003001

Taylor, P.G., Upham, P., McDowall, W., Christopherson, D., 2014. Energy model, boundary object and societal lens: 35 years of the MARKAL model in the UK. Energy Research and Social Science 4, 32–41. https://doi.org/10.1016/j.erss.2014.08.007

UCL, 2021. UK MARKAL [WWW Document]. URL https://www.ucl.ac.uk/energy-models/models/uk-markal (accessed 5.30.21).

UKERC, 2009. Making the transition to a secure and low carbon energy system: synthesis report of the Energy 2050 project. UKERC, London.

UKERC Energy Data Centre, 2021. Centre for Energy Systems Integration [WWW Document]. URL https://ukerc.rl.ac.uk/cgi-bin/ercri5.

pl?GChoose=gregsum&GRN=EP/P001173/1&GrantRegion=10&GrantOr
g=109&HTC=361DDE2&SHTC=80680D (accessed 5.30.21).

van der Vleuten, E., 2004. Infrastructures and societal change. A view from the large technical systems field. Technology Analysis and Strategic Management 16, 395-414. https://doi.org/10.1080/0953732042000251160

Wheatcroft, E., Wynn, H., Dent, C.J., Smith, J.Q., Copeland, C.L., Ralph, D., Zachary, S., 2019. The Scenario Culture. https://arxiv.org/abs/1911.13170.

Winskel, M., Ketsopoulou Irina, Churchhouse, T., 2015. UKERC interdisciplinary review. UKERC, London.

CHAPTER 3

Environment-Friendly Energy Research in Norway

Abstract The Research Council of Norway established the Centres for Environment-Friendly Energy Research in 2009. These are long-term national centres that are meant to integrate academics with industries, private companies, regulating bodies, governmental organisations, and research institutes, to trigger a clean-energy transition and pursue environmental innovations. Increasingly, addressing energy issues through the integration of technological and Social Sciences and Humanities disciplines has become expected in these Centres. This chapter draws from interviews with the project participants and fieldwork to demonstrate how different academics and professionals experienced these interdisciplinary collaborations, including what consequences and dynamics such collaborations generated. We round up by interpreting the findings along with the traits of interdisciplinarity that have been emerging in this book.

Keywords The Research Council of Norway • Research funding • Publications • Social Sciences and Humanities • Environmental innovation

© The Author(s) 2022
A. Silvast, C. Foulds, *Sociology of Interdisciplinarity*,
https://doi.org/10.1007/978-3-030-88455-0_3

49

3.1   Introduction

Low-carbon transitions of energy systems are multi-dimensional and complex sociotechnical processes. As has been now shown in this book, it has become increasingly expected that interdisciplinary research—that crosses academic disciplines, as well as quite often sectors and policy areas too—will develop new knowledge and help address this situation. Over the past years, academic disciplines from the Social Sciences and Humanities (SSH) have therefore been especially expected to contribute new knowledge to energy research, policy, and innovation and hence further energy transitions in societies.

In line with this trend, the Research Council of Norway established the Centres for Environment-friendly Energy Research (Forskningssentre for miljøvennlig energi, from here on, FMEs, as they are commonly abbreviated) in 2009. These are long-term national centres that are meant to integrate academics with industries, private companies, regulating bodies, governmental organisations, and research institutes to trigger a clean-energy transition and pursue environmental innovations. Increasingly, integrating technological and SSH disciplines to address energy issues has become normatively required in these Centres. The chapter draws from interviews with the project participants and fieldwork to demonstrate how different academics and professionals worked in these interdisciplinary collaborations and what consequences and dynamics their knowledge production has had. In Chap. 5, we move to discuss on a more general level how a consultancy evaluated these large-scale interdisciplinary Centres.

This chapter presents findings from fieldwork concerning the FMEs, conducted in 2019. Science and Technology Studies (STS) has led us to consider the FMEs not just as a new policy instrument, but also as an opportunity for conducting field research on how its research was actually carried out in a collaboration (with the SSH, as well as other disciplines). The corpus includes qualitative interviews with professionals aligned with and around the FMEs. The scope of the data collection spans various actors including research group leaders, coordinators, and researchers. Ten interviews have been conducted with representatives of seven different FMEs. Although not a large number and not representative of all FMEs, these data are complemented by other means and offer a unique view into the knowledge production and its dynamics happening within the FMEs.

In contrast to the classic objects of Science Studies (cf. Silvast and Virtanen 2019), the FMEs are primarily not set up as laboratories, research groups, or other single sites of expert knowledge. Some of them have an actual office where they are located and that could be visited by an ethnographer, but some of them do not and they exist mainly as virtual-networked organisations. The FMEs also run seminars, workshops, training for PhD researchers, and other related events, but these do not happen all the time and will involve various subsets of the FME members. As large centres, their trait is being geographically distributed across Norway and they were initially conceived to set up 'national teams'. They involve very different kinds of actors, from researchers in various universities and research institutes to project partners in the industry, business, and public enterprises that also provide about half of the FME funding (Impello 2018). All of this means difficulties for conducting a 'field study' of the FMEs because of the difficulty of defining 'what a field is' in this case.

To address this methodological requirement of overcoming the single-case research designs, this research uses several strategies. The first is studying many of the FMEs in the same (cross-FME) projects, rather than just a single case study in one of them. Secondly, we use other techniques from interviewing to place the FMEs in the broader context of research policy and the evolution of academic research in Norway. To accomplish this, the research presented in this chapter gathered an overview of what has been written about the FMEs often by their participants themselves by developing an SSH perspective. A literature review of the FMEs was completed and is depicted in Table 3.1. A selected subset of all the discovered

Table 3.1 Reviewing the Centres for Environment-Friendly Energy Research in scientific databases

Search term	Oria[a]	Web of Science	Scopus	Google Scholar
"Centres for Environment-Friendly Energy Research"	9	3	4	382
"Forskningssentre for miljøvennlig energi"	9	0	0	80
"Forskningssentrene for miljøvennlig energi"	1	0	0	24

[a]Oria is a unified library database in Norway, including books, articles, magazines, music, films, and electronic resources. It is maintained by BIBSYS, an agency established by the Ministry of Education and Research in Norway. It is a collaboration among all the Norwegian universities, university colleges, research institutions, and the National Library of Norway. The Norwegian University of Science and Technology (NTNU) has the formal organisation of BIBSYS

studies was gathered, focusing on those works that apply SSH (whereby the Social Sciences includes Economics) perspectives or methods to examine the FMEs or that embed energy issues in economy or society whilst discussing the FMEs. The result was 75 works developed from a socio-technical perspective, including both peer-reviewed journal articles and grey papers. In what follows, both the interviews and desk-based overviews of this literature are deployed to address the main research goals.

This chapter is structured as follows: we begin with background context on the FMEs themselves (Sect. 3.2), after which we summarise the outcomes of our analysis on FME publication outputs (Sect. 3.3.1). Following this context, we first examine how interdisciplinary knowledge production is carried out in FMEs (Sect. 3.3.2); qualitatively map the academic disciplines in the FMEs (Sect. 3.3.3); and study the interpretations of innovation that the FMEs are drawing upon (Sect. 3.3.4). We finish this chapter with some conclusions on a programme at the Norwegian University of Science and Technology (NTNU) that has tried to straddle all the FMEs and combine them under one 'umbrella' initiative on energy transitions, asking what such merger means for interdisciplinarity.

3.2   What Is Environment-Friendly Energy Research and Innovation in Norway?

The political consensus on addressing climate issues and the role of academic research in generating knowledge to that aim are long-standing developments in Norway. The Norwegian Declaration of Soria Maria in 2005 emphasised environment and climate and stressed global challenges including energy issues. The 2008 climate contract by all political parties increased research resources on clean energy studies (Pelkonen et al. 2010). Energi21, a national strategy for research, development, and commercialisation of climate-friendly energy technologies, was established in the same year. Evaluations have argued that SSH research on energy and climate issues was prioritised in these programmes (Klitkou et al. 2010). They also claimed that had been a visible growth in research collaborations, networks, and scientific articles and citations in Norwegian energy research that draws on SSH perspectives (Ramberg et al. 2016).

In practice, however, the contributions of and tasks attributed to SSH in environment-friendly energy research varied across research programmes and changed substantively over time. In 2009, the Research

Council of Norway established a historical response to the climate agreement by granting eight Norwegian Centres for Environment-Friendly Energy Research (Forskningssentre for miljøvennlig energi, FME). Characterised initially as a technology-push initiative (Jakobsen et al. 2019), the FMEs are long-term centres whose remits include increasing innovation; contributing to national and international emissions reduction, energy efficiency, and renewable energy; promoting the development of research environments; and knowledge-based contributions to energy debates (Norges forskningsråd 2018a). The FMEs work through university-industry collaboration (Nilsen and Lauvås 2018) and focus on what researchers term as environmental innovations. These are products, production processes, services, and management and business models that are novel in an organisation and reduce negative impacts to the environment (Jakobsen et al. 2019).

These initial FMEs focused on a number of technological areas: including energy use in buildings and regions; solar energy and materials for solar cells; bioenergy including biofuels; hydropower; Carbon Capture and Storage; and energy systems. Between now and then, a large share of the FME funding has been distributed to research on buildings, solar energy, and bioenergy (Impello 2018).

While all the initial Centres were technological, three further were granted in 2011 specifically on SSH aspects of energy and climate (CICEP, Strategic Challenges in International Climate and Energy Policy; CREE, Oslo Centre for Research on Environmentally Friendly Energy; and CenSES, Centre for Sustainable Energy Studies). These SSH-based FMEs aimed to improve the knowledge base for energy policy and public and private decision making (Government.no 2013). The year 2016 saw the granting of further, both technological and SSH-focused FMEs which included new themes on energy efficiency in industry and zero emission transport. In 2019, the Research Council of Norway granted two more SSH-focused FMEs on energy transition strategies and socially inclusive decarbonisation, respectively. The newest FME to date is the Norwegian Research Centre on Wind Energy. An overview of the areas of the FMEs to date is in Table 3.2.

At this point, the integration of SSH disciplines also became mandatory for the technological Centres. As the Research Council of Norway (Norges forskningsråd 2018b, p. 2) summarised, the FME scheme is meant to integrate key disciplines and research environments related to energy issues: it "engages social scientific, humanities, and science and technology

Table 3.2 Overview of the Centres for Environment-Friendly Energy Research, as of 15 June 2021. Data sourced from project database of the Research Council of Norway

Name	Budget (M)	Project period
Technology-oriented FME centres		
Norwegian Research Centre on Wind Energy	NOK 120	2021–2029
Research Centre for Sustainable Solar Cell Technology	NOK 115.6	2017–2024
Norwegian Centre for Sustainable Bio-Based Fuels and Energy	NOK 124.8	2017–2024
Mobility Zero Emission Energy Systems	NOK 120	2017–2024
Centre for an Energy Efficient and Competitive Industry for the Future	NOK 200.2	2016–2024
The Research Centre on Zero Emission Neighbourhoods in Smart Cities	NOK 176	2016–2024
Norwegian CCS Research Centre	NOK 184.4	2016–2024
Norwegian Research Centre for Hydropower Technology	NOK 192	2016–2024
Centre for Intelligent Electricity Distribution	NOK 160	2016–2024
Subsurface CO2 Strategy	NOK 80	2010–2018
FME Solar United	NOK 160	2009–2017
Norwegian Centre for Offshore Wind Energy	NOK 120	2009–2017
FME CenBIO (Bioenergy Innovation Centre)	NOK 120	2009–2017
Centre for Environmental Design of Renewable Energy	NOK 80	2009–2017
The Research Centre on Zero Emission Buildings	NOK 120	2009–2017
BIGCCS Centre (International CCS Research Centre)	NOK 160	2009–2017
FME NOWITECH (Research Centre for Offshore Wind Technology)	NOK 160	2009–2018
Social Sciences-related FME centres		
FME INCLUDE (INCLUsive Decarbonisation and Energy Transition)	NOK 95	2019–2027
Norwegian Centre for Energy Transition Strategies	NOK 105	2019–2027
CICEP (Strategic Challenges in International Climate and Energy Policy)	NOK 64	2011–2020
CREE (Oslo Centre for Research on Environmentally Friendly Energy)	NOK 64	2011–2020
CenSES (Centre for Sustainable Energy Studies)	NOK 80	2011–2019

See: https://prosjektbanken.forskningsradet.no/explore/projects?Kilde=FORISS&distribution=Ar&chart=bar&calcType=funding&Sprak=no&sortBy=date&sortOrder=desc&resultCount=30&offset=0&ProgAkt.3=FMETEKN-FME+-+teknologi and https://prosjektbanken.forskningsradet.no/explore/projects?Kilde=FORISS&distribution=Ar&chart=bar&calcType=funding&Sprak=no&sortBy=date&sortOrder=desc&resultCount=30&offset=0&ProgAkt.3=FMESAMFUNN-FME+-+samfunn

research environments, and in several cases, takes an interdisciplinary and multidisciplinary approach".

Explicitly, the FMEs are expected to engage the SSH in the study of energy policy, behavioural change, dissemination of new technologies, and business studies (Norges forskningsråd 2018b, p. 8). These are typical, anticipated contributions from the SSH disciplines in energy research (Mallaband et al. 2017; Silvast et al. 2013). They appear at a partial stage of innovation: they contribute new knowledge and address innovations at their very late stage, of diffusion. Typically, this means that SSH are requested to study the behaviour of eventual technology 'end-users' (Ingeborgrud et al. 2020; Silvast et al. 2018) or public acceptance of new technologies (Robison and Foulds 2021; Ryghaug et al. 2018). This is where strong SSH contributions are envisaged, essentially to overcome resistance: it has been argued that the scientist participants in one of these FME centres anticipate a public that lacks information and will be resistant to new energy technologies (Heidenreich 2015).

However, emerging international research has spoken to considerably wider participation of different academic disciplines and stakeholders in the earliest stages of innovation and technology development. This is meant to contribute to the anticipation, reflection, engagement, and activity of all relevant societal actors in technology design, that is to say, Responsible Research and Innovation (RRI) (Rommetveit et al. 2017; von Schomberg 2011). Especially sustainable development has been associated with a wholly new concept of Transformative Innovation Policy (TIP) (Schot et al. 2018; Schot and Steinmueller 2018). This concept designates innovation that goes beyond technology development and design and embraces inclusivity, organisational change, and experimentation. The transformative focus implies acknowledging "civil society and citizens as not only consumers and adopters of innovation but as promotors and sources for innovations which address social and environmental needs" (Schot et al. 2018, p. 8).

The FMEs, with their primary focus on environmental innovation, line up with these considerations directly. Several detailed studies on the working practices of FMEs have been conducted by academics, examining how their members describe trade-offs between innovations and academic outputs (Nilsen and Lauvås 2018) and respond to external energy policy objectives coherently (Åm 2015; Jakobsen et al. 2019). But also, in contrast to several countries with a long-term innovation model—such as Finland, Sweden, and the Netherlands—the first Norwegian official joint statement on innovations was published relatively recently in 2008 (Norwegian Ministry of Trade and Industry 2008). More recently, the discussion has expanded in the policy, professional, and academic sense, and the Research Council of Norway has backed both the RRI and the TIP concept (TIP Consortium 2017). As these considerations demonstrate, the role of SSH in the FMEs and environmental innovation merits further interrogation and has not been fully acknowledged by existing research and evaluations.

3.3   The Role of Social Sciences and Humanities in Norwegian Environment-Friendly Energy Research and Innovation

3.3.1   Centres for Environment-Friendly Energy Research in the Literature

The collected publications on the FMEs, deploying SSH perspectives (Fig. 3.1), display a clear trend. Almost directly after their foundation in 2009, there is a rising interest in studies of FMEs. This is understandable:

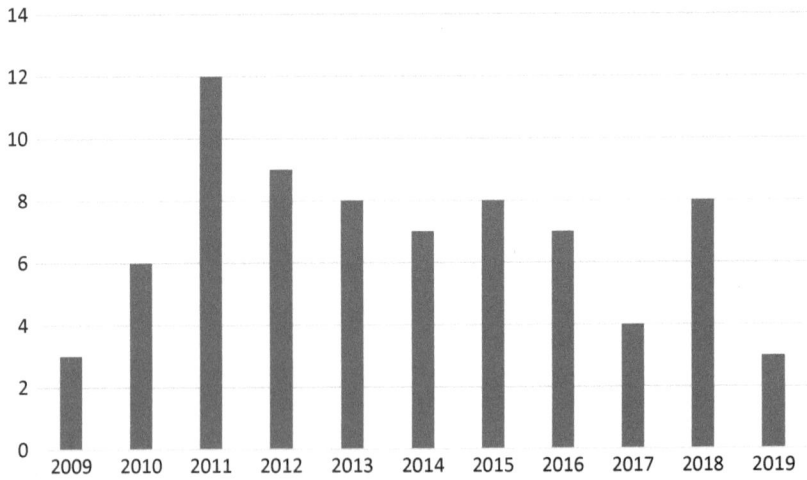

Fig. 3.1   The number of Social Sciences and Humanities publications published by all Centres for Environment-Friendly Energy Research, per year (2009–2019). Sources: Oria, Web of Science, Scopus, Google Scholar

an increasing number of works will find it interesting to study that there is a new major funding instrument in Norway. After 2011, the number drops off and there is no visible trend after that, which suggests that while the number of FMEs had increased, there was no consistent accumulation of discussions on them. The publication types (Fig. 3.2) show that the majority of this type of publishing has happened in various grey literatures, especially reports and non-reviewed conference papers. This can be readily explained by the composition of these Centres: some of the FMEs are led by and have large budgets in public research institutes that do not always have a strong academic publishing tradition. Journal articles, however, represent the second largest group. The articles are not published in any particular main journal, with the exception of *Energy Procedia* that publishes conference proceedings and contains the largest number of publications. The outlets of the rest range from the Norwegian *Forskningspolitikk* (research policy) to *Environmental Innovation and Societal Transitions, Economic Geography, Energy Policy*, and *Science and Public Policy*, and various others with only a few publications per each journal.

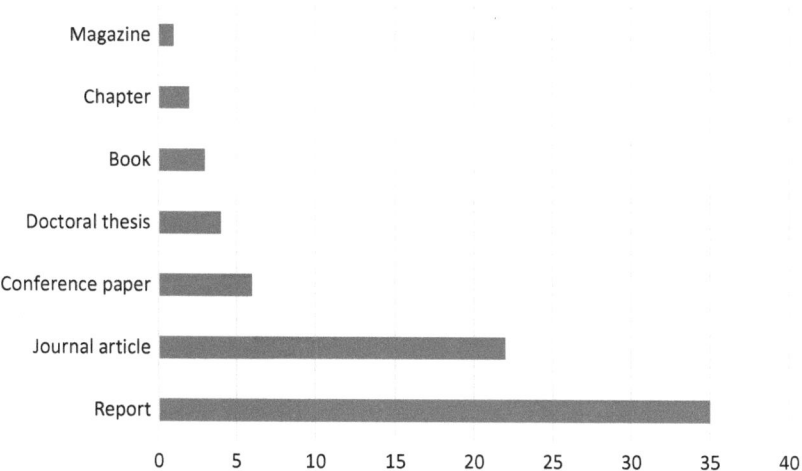

Fig. 3.2 The number of Social Sciences and Humanities publications published by all Centres for Environment-Friendly Energy Research, per publication type (2009–2019). Sources: Oria, Web of Science, Scopus, Google Scholar

The author-provided keywords provide an important first viewpoint into what the scholars themselves were writing about, when reflecting on the FMEs from a sociotechnical perspective. The complete list of words is in Table 3.3. One can immediately see the lack of important keywords: there is no interdisciplinarity, multidisciplinarity, or transdisciplinarity. Even more strikingly, there are no keywords on any kind of innovation. There are, however, a number of visible interests in policy and politics including both energy and climate. The links between the industry and

Table 3.3 Authors' keywords when studying the Centres for Environment-Friendly Energy Research

Keyword
• Absorptive capacity
• Actor-based approach
• Agenda setting
• Arena of development
• Branching
• CCS
• Climate policy
• Emissions
• Energy
• Energy policy
• Environment
• Environmental Sciences and Ecology
• Renewables
• International R&D consortium collaboration
• Climate gas
• Knowledge integration
• Learning curve
• Norway
• Offshore wind
• Path creation
• Path dependence
• Path interdependencies
• Policy
• Politics
• R&D support
• Solar
• Sustainability transitions
• Technological development
• Translation

Sources: Oria, Web of Science, Scopus, Google Scholar

the academia—such as research and development (R&D) issues—also clearly received interest. Several articles are written centred on technologies as such, corresponding with how the FMEs were setup: including offshore wind and solar energy. That said, there are also some more systems-oriented perspectives, which mainly seem to stem from various kinds of transitions studies. These include focuses on path dependencies and path creation, path interdependencies, and sustainability transitions as such.

The interpretation of these key focuses follows from the way in which the SSH were meant to work in the FMEs, by design. As mentioned above, the funder's interest in "social science FMEs" corresponded with a focus on energy policy, decision making (Government.no 2013), behavioural change, dissemination of new technologies, and business studies (Norges forskningsråd 2018b, p. 8). With the exception of behavioural studies, nearly all of these themes are present in the keywords. When publishing about the FMEs, SSH perspectives do not seem to have started from transgressing disciplinary boundaries, but closer to their set task, which was providing the FMEs with a social and political framing and increasing understanding of how the FMEs could collaborate between the academia and the industries. It is also revealing that these assumed SSH perspectives did not address innovation as their topic. This may be because innovation did not feature as a central concept or frame in the research practice and the research outputs; or because when innovations were defined, they were framed simply as commercialised inventions (Schot and Steinmueller 2018) to whose study many disciplines from the SSH did not have a predefined role. This finding will be deepened by the interviews with the FME project experts.

3.3.2 Interdisciplinary Knowledge Production

The following draws from the completed interviews for this research, gathered from different seniorities in various FMEs in 2019. These informants that had experienced working in the FMEs recognised that interdisciplinarity had been a strong requirement when the FMEs were setup, especially since 2011 after the initial rounds that had centred on technological disciplines. As many of them continued, a difference exists, however, between designing interdisciplinary collaboration as part of an application and making this collaboration happen in everyday research and all the time. These are well-recognised issues in the interdisciplinarity studies (Winskel 2018; Winskel et al. 2015) and were suggested in the Norwegian case. Firstly, integrating different knowledge from various disciplines takes time and effort that were not always sufficiently anticipated by the participants or in research designs. Had this social learning been allocated more resources, it might have happened more frequently or rapidly according to the research participants. In the early centres, it happened often but in certain situations, as when one of the early Centre project members used social practice theory and the Multi-Level Perspective to translate some SSH insights to building research practice:

I think I used [sociologist and practice theorist Elizabeth] Shove's work a lot, Comfort, Cleanliness, and Convenience. That was very easy to communicate [for building managers and engineers]. At some, we also used a lot of the Multi-Level Perspective (MLP). Despite all its criticisms, it is really easy to communicate, it opens up some ice. [At] One of these conferences, I used the MLP to contrast the diffusion curve with the MLP. I think it was useful for many people. (Man, 50–59 years, Professor)

This quote points to the role of reducing complexity and translations, even when the concepts to do this, such as the MLP, can be criticised from certain perspectives. This should not overlook, however that secondly, knowledge from various academic disciplines is not always only difficult, but can in some cases be nearly impossible to integrate, even as a purely practical matter. A typical example, which was pertinent to many of the Norwegian Centres, was the difference between the modelling-based knowledge used for example in optimising power systems or analysing sensor data from buildings, as compared to the more qualitative evidence expected from the SSH. When these kinds of knowledge practices meet, they reveal different epistemic values, assumptions, and units of analysis that cannot be reconciled by simply, for example, feeding social scientific data into energy systems models. Thirdly, the informants recognised that academia's reward structures are not ideally suited for interdisciplinary knowledge production—as, for example, many key scientific journals in research fields remain monodisciplinary. It may be telling of this that many of the FMEs, when they published on SSH issues, did it on channels that were not academic journals (see the previous section).

While working across disciplines was often seen as in need of continuous maintenance, this did not mean that SSH had no role in the research practices. In many cases, the SSH were given a predefined task in the

Centres whether by work package structure or set roles on what these disciplines were expected to do. In the example of research on energy efficiency in buildings and sustainable energy studies, the typically assigned role was studying the 'users', whether in user experiments or in behaviour of the resident 'users' or households more generally. This assumption also brought about a specific temporal location in the research process—the SSH were often introduced to the studies 'after the fact' when the technologies had already been designed and needed to be diffused to their expected 'users'.

SSH perspectives could also find roles in some of the other key concerns of the FMEs. One of these was innovation, as two scholars who had both worked in FMEs and researched them explained. There was a different agreement on what it would mean to be an innovation scholar that shaped how the end-user partners and the SSH researchers would expect to work in the projects:

> A: Some [user partners] think that OK, you are an innovation researcher, so you are the ones that are going to make the innovation happen, that is the most extreme view that you meet. (Man, 30–39 years, Researcher)

> B: We were responsible for our innovation, that we manage and create innovation, commercialisation, technology, and so on. We really have to tell them real hard that we are doing research and that research has implications for how they manage innovation. (Woman, 40–49 years, Research Director)

> A: that we can contribute to the process, but they cannot "outsource" it to the social sciences, they have to do the technical development. We have done work for one of the FMEs and given input to the centre management how it is going, and what they can do differently. (Man, 30–39 years, Researcher)

This quotation shows both the differing expectations—where the user partners assumed that SSH scholars would make innovations happen, the SSH scholars thought they would do research that would frame innovations for the partners—and the clear need for active maintenance work in interdisciplinarity. The SSH scholars needed to do continuous maintenance of their relationship with others and express what their contribution to knowledge is. In the term 'outsourcing', there is an implicit assumption that was the relationship not actively maintained, it would default back to kind of a provider of a service, such as innovation management instead of scholarship of innovation in its own right.

There were several other similar examples of roles and expectations. Another potential role again related to the interest of several industrial FME user partners, namely communication. Communication meant making the results of the FMEs more visible outside of the circle of their developers. In making this communication happen, it seems to have been recognised that 'social' framing, as well as 'technical' framing, of the new technologies was necessary. But it would be premature to associate this communicative practice with inclusive attention to innovations. This is because it operated mainly to one direction. When it involved other actors than developers and designers, it did so simply aim at informing these of new scientific discoveries.

3.3.3   Academic Disciplines Involved from Social Sciences and Humanities

The FMEs are technology-oriented, but in planning them, crossing disciplines has been increasingly required and the SSH have been preferred to be involved. Disciplinary diversity was hence prized by the FMEs and their participants. That said, when attempting to map which disciplines exactly were involved in these collaborations, it is important to question the relevance of disciplinary labels especially as part of everyday working in the Centres. To some extent, the interviewees drew on disciplinary 'gaps' by using the labels when we mentioned them. Generally, there simply was not that much SSH involvement, even in terms of number of staff, in the more technological Centres. This also meant lacking some disciplines for the problems that these Centres were addressing. For example, Environmental Psychology could have brought insights on buildings and energy efficiency, and planning-oriented disciplines—such as Geography or Sociology—might have been of considerable help in researching energy in urban areas. One participant from an SSH-focused FME also made distinctions between which disciplines from the SSH integrated more than the others. She had found out that disciplines studying 'technology users' had clear contributions, but that the Political Sciences had found their roles more contested, on the count of introducing findings on the political implications of energy technologies which some participants might have not preferred as objects of analysis.

Yet, it was also clear that these disciplinary labels were somewhat constructed in the interview situation itself. It did not seem like everyday research in the FMEs had relied on scholars always identifying with their own academic discipline, for example, based on where they received their doctoral degree. The case of 'user studies' is again a good illustration, as this problem could be engaged by a number of disciplines and interdisciplinary fields, ranging from STS to Sociology, Anthropology, Human-Centred Design, Cultural Studies, and beyond that. Scholars who pursue understanding the 'user' may do so in terms of that practical problem rather than as anthropologists, sociologists, and so forth.

Another popular example, related to installing large-scale energy technologies—such as Norwegian hydropower—in localities, is when the role of SSH becomes translated into 'social acceptance' research. An interview quote below makes no practical difference between the Social Sciences and social acceptance research, but, for this purpose, equates them with one another:

> When it comes to the social sciences, the companies are also in the process of maturing. Because there's been a lot of focus on the biological part, hydropower companies have been addressing fish populations ... What they see now is to include and look at the social acceptance, when it comes to upgrades and expansion, they have to have a kind of addressing and some kind of insights into social acceptance issues. A very good acceptance and acknowledging the social sciences in order to have a good collaboration with local interests. (Woman, 40–49 years, Executive Director)

She worked for one of the technical FME centres, and one could assume that with the link from energy installations to social acceptance, local interests, and the Social Sciences to study and translate between interests and acceptance them, there would be a very large role for crucial SSH insights both conceptually and in the applied sense. Indeed, this is a large and evolving research area, where new openings exist spanning from Political Science to Anthropology, Sociology, Geography, Ecology, and beyond (Wolsink 2018). To do research on users, social acceptance, or other themes framed by technologists does not automatically cause restricted research agendas.

But this was still not always the case, according to the participants. An important and critical example of disciplinary positions and roles needs to be explained here—this one used by the interviewed SSH scholars themselves. Some of the project participants felt that their SSH perspectives had been minor in the research. This could have happened in a number of different ways. One of them directly followed from the social acceptance focus and was based on the deficit model of the public (Ryghaug et al. 2018)—namely, in another Centre than the one above, it was first assumed that SSH study the 'end-users' and acceptance issues, but then questioned whether it made sense to interview these technology users when they might have lacked relevant information on energy technologies and hence been unable to generate new scientific knowledge on these technologies. Another assumption at play was associating SSH with mere theory but not connected to actual development and design of technology. These assumptions on ineffective contributions by these disciplines were sometimes reinforced by project design, for example, by letting the SSH participants sit in management boards but keeping them at a distance as one interview participant (Man, 40–49 years, researcher) put it, from everyday research practice.

3.3.4   The Interpretations of Innovation

Considering that the FMEs were explicitly set up to increase innovations, it is important to highlight how much the concept of innovation had fluctuated during the life course of these large Centres. Some of the informants suggested that innovation was a new concept to the FMEs themselves. In fact, we should stress the unusual place of these centres in the Norwegian funding system, since innovation funding and research funding had historically separate funding streams, expectations, and even institutions that deal with each of them (with, for instance, SINTEF focusing on innovation and NTNU focusing on research, though the two collaborate intensely). One of the participants in the early centres noted:

> In 2009, the FMEs were new, and research institutions had to find out how to organise them. And we discussed this after 4 years in one of the FMEs that this is a hybrid: between centre of excellence research and a centre for innovation, which are two funding instruments that had existed before. It was supposed to do both at the same time, producing high-quality research and also producing practical solutions to climate change. (Man, 50–59 years, Professor)

The concept of innovation had strengthened after the evaluations of the FMEs, when some industrial partners wanted to seek more problem-oriented solutions from the Centres. Accordingly, the industrial partners expected short-term and concrete results, which some interviewees termed directly as *innovations*.

Around this same time, the funders seem to have systematised their attention on innovation. But this initial work linked with pursuing the Centres to use the concept of innovation at all, rather than more reflective or theoretical discussions on what the concept could mean. Many stressed that innovations were mainly seen as commercialised inventions that should ideally be counted and otherwise measured. The key tool of this calculation of produced activity were Technology Readiness Levels: an estimate on the maturity of technologies, especially in this case, to enter the markets as new products. According to many of the accounts, the recent discussion in FMEs on innovations centred on how the technologies developed could climb the Technology Readiness Levels towards actual demonstrations and products. As this was still a relatively new discussion, it had to be pursued by special arrangements at the time of the study, such as involving professional 'innovation managers' in the Centres and setting up innovation boards where they would meet the other project members and leaders.

Innovation managers and other related professionals were not clearly experts of any particular academic discipline, although their work leaned on commercial principles. That said, one project participant accounted an earlier experience, where the SSH were designed into the Centre to act as innovation experts in the management groups. These arrangements point out to the flexibility of the idea of innovation and how it opened up to participation of, at least, various kinds of experts in the academia and those mediating in between universities and the industries.

3.4   Conclusions

The chapter addressed how the FMEs have worked as a way of conceptualising transformative change anew in interdisciplinary energy research and what relationships to policy they were meant to develop. The findings show how the energy transformation in these cases was opened up to the broad involvement of different academic disciplines, including SSH, but also limited to this involvement of each of these disciplines to specific set roles. The result was not radical interdisciplinarity per se, but instead academic disciplines pursuing research on a predefined task, whether it be 'end-users' or energy policy in the case of SSH. In fact, the high-level labels set around 'social scientific' and 'technological' FMEs by the funding body suggest two dynamics, already recognised in the previous chapter. First, there is an indication of SSH being appropriated (Forsythe

1999) by other disciplines by configuring them certain roles in the project. Second, the fact that some FMEs are still called Social Science suggests that the importance of disciplines has not diminished in these interdisciplinary collaborations. On the contrary, the SSH are brought forward as a seemingly unified one discipline that can address relevant energy issues. There are grounds to draw on such labels—a theme we return in Chap. 5—but it needs to be done strategically, while respecting the diversity of SSH disciplines and highlighting the often-backgrounded techniques and tools that being labelled as a certain kind of 'social scientist' (or SSH scholar) can offer to other academics.

The research also discussed how the concepts of environmental innovation and university-industry collaboration were framed in the context of the FMEs and how the FMEs themselves have perceived these new approaches. These new approaches involved industrial partners, public institutions, and various others, but it was also visible who was not involved: there was little evidence of civil society actors, NGOs, or social movements being engaged in the FMEs' work and innovations that were studied. The concepts of environmental innovation and transdisciplinary collaboration have evidence of interpretative flexibility. We can critique these different interpretations for what they miss, but on the other way, they also likely have produced coherence to the projects studied. In this, they can function as 'boundary objects' (Star and Griesemer 1989) that integrate different social worlds of the project and give them goals to aspire to.

The role of research funding becomes perhaps most pronounced in this chapter, as it has set about explicitly studying large national centres that were established and financed by the national funding organisation. It is clear that the funder has been actively discussing this interest and developing new viewpoints on what the Centres were achieving (Impello 2018). We revisit this theme in depth in Chap. 5, when drawing together our interest on the impacts of research funding on interdisciplinary working.

We now conclude this chapter by drawing these interests together in the context of the Norwegian University of Science and Technology's Energy Transition Initiative (NETI), a research programme established at the NTNU between the main energy industry players and researchers. NETI aimed at generating knowledge-based energy transitions and setting up research environments to pursue this knowledge; one of the potential functions to do this was for it to become the 'umbrella organisation' of the FMEs in Norway. As it were, NETI wanted to stand as a hub to all the FMEs and their research problems and areas. Aside from the strategic aims of the initiative, it is in the interest of this book because it demonstrates whether, and if so how, the current energy transition can be merged under one overarching approach especially as different academic disciplines are pursuing knowledge on it. The NETI offered to do this by not offering one single strategical pathway, but multiple strategies to transition in future energy systems. Its main themes, sourced from internal presentations (e.g. see Silvast 2019), were as follows: energy policy and scenarios; innovation and entrepreneurship; human behaviour/consumer research; energy storage, distribution, and technology development; energy markets and business models; and sustainability/climate research.

These themes are revealing of the diversity of issues and problems that an energy transition would require, according to the characterisation of NETI. The interviews in this chapter have similarly demonstrated the diversity of these topics, including the various epistemic cultures (Knorr Cetina 1999) at play, and how social learning was required in the FMEs to effectively cross between them.

It is also important to recognise that these themes around transition are organised around topics and problems, rather than broad academic disciplines. This problem-orientation poses a relevant approach to transition strategies but excludes the particular tools and methods that might be used to increase, for example, innovation, sustainability, and understanding in consumer research. In this way, the image masks epistemic differences between disciplinary tools—such as computer models used in scenarios and optimisation, as opposed to psychological, sociological, and anthropological knowledge on people and use of technologies. In order for a transformative innovation to be pursued, these epistemic values of different tools would require further attention. The orientation to problems as opposed to disciplines also poses a further problem: that many fields, such as STS, would not study these problems in isolation from one another, as they are quite clearly interrelated; for instance, energy policy is closely related to energy market design (Silvast 2017). In summary, the problem-oriented interdisciplinary research does not offer a single solution to the problems of disciplinary-based academic knowledge production.

References

Åm, H., 2015. The sun also rises in Norway: Solar scientists as transition actors. Environmental Innovation and Societal Transitions 16, 142–153. https://doi.org/10.1016/j.eist.2015.01.002

Forsythe, D.E., 1999. "It's just a matter of common sense": Ethnography as invisible work. Computer Supported Cooperative Work 8, 127–145. https://doi.org/10.1023/A:1008692231284

Government.no, 2013. Centres for environment-friendly energy research [WWW Document]. URL https://www.regjeringen.no/en/topics/energy/energy-and-petroleum-research/centres-for-environment-friendly-energy-/id633931/ (accessed 5.24.21).

Heidenreich, S., 2015. Sublime technology and object of fear: Offshore wind scientists assessing publics. Environment and Planning A 47, 1047–1062. https://doi.org/10.1177/0308518X15592311

Impello, 2018. Effekter av energiforskningen. [The effects of energy research.] Impello, Trondheim.

Ingeborgrud, L., Heidenreich, S., Ryghaug, M., Skjølsvold, T.M., Foulds, C., Robison, R., Buchmann, K., Mourik, R., 2020. Expanding the scope and implications of energy research: A guide to key themes and concepts from the Social Sciences and Humanities. Energy Research and Social Science 63, 101398. https://doi.org/10.1016/j.erss.2019.101398

Jakobsen, S., Lauvås, T.A., Steinmo, M., 2019. Collaborative dynamics in environmental R&D alliances. Journal of Cleaner Production 212, 950-959. https://doi.org/10.1016/j.jclepro.2018.11.285

Klitkou, A., Pedersen, T.E., Schwach, V., Scordato, L., 2010. Social science research on energy International and Norwegian studies. NIFU, Oslo.

Knorr Cetina, K., 1999. Epistemic cultures. Harvard University Press, Cambridge, MA. https://doi.org/10.2307/j.ctvxw3q7f

Mallaband, B., Wood, G., Buchanan, K., Staddon, S., Mogles, N.M., Gabe-Thomas, E., 2017. The reality of cross-disciplinary energy research in the United Kingdom: A social science perspective. Energy Research and Social Science 25, 9-18. https://doi.org/10.1016/j.erss.2016.11.001

Nilsen, T., Lauvås, T.A., 2018. The role of proximity dimensions in facilitating university-industry collaboration in peripheral regions: Insights from a comparative case study in Northern Norway. Arctic Review on Law and Politics 9, 312-331. https://doi.org/10.23865/arctic.v9.1378

Norges forskningsråd, 2018a. Forskningssentrene for miljøvennlig energi (FME): Resultater og høydepunkter fra åtte FME-er. [Research Centers for Environmentally Friendly Energy (FME): Results and highlights from eight FMEs.] Forskningsråd, Oslo.

Norges forskningsråd, 2018b. Energi (ENERGIX, FME, CLIMIT) årsrapport 2018. [Energy (ENERGIX, FME, CLIMIT) Annual Report 2018.] Forskningsråd, Oslo.

Norwegian Ministry of Trade and Industry, 2008. An innovative and sustainable Norway. Oslo.

Pelkonen, A., Teräväinen, T., Häyrinen-Alestalo, M., Waltari, S.-T., Tuominen, T., 2010. Tiedepolitiikan kansainvälisiä kehitystrendejä 2000-luvulla. [The International Trends of Science Policy in the 2000s.] Opetus-ja kulttuuriministeriön julkaisuja, Helsinki.

Ramberg, I., Børing, P., Klitkou, A., Solberg, E., 2016. Social science research on environmentally friendly energy in Norway. NIFU, Oslo.

Robison, R., Foulds, C., 2021. Social sciences and humanities for the European green deal: 10 recommendations from the EU Energy SSH Innovation Forum. Cambridge.

Rommetveit, K., Dunajcsik, M., Tanas, A., Silvast, A., Gunnarsdóttir, K., 2017. CANDID PRIMER: Including Social Sciences and Humanities scholarship in the making and use of smart ICT technologies. CANDID research project paper. University of Bergen, Bergen.

Ryghaug, M., Skjølsvold, T.M., Heidenreich, S., 2018. Creating energy citizenship through material participation. Social Studies of Science 48, 283–303. https://doi.org/10.1177/0306312718770286

Schomberg, R. von, 2011. Towards responsible research and innovation in the information and communication technologies and security technologies fields. Publications Office of the EU, Brussels.

Schot, J., Boni, A., Ramirez, M., Steward, F., 2018. Addressing the sustainable development goals through transformative innovation policy. TIP Consortium, Brighton.

Schot, J., Steinmueller, W.E., 2018. Three frames for innovation policy: R&D, systems of innovation and transformative change. Research Policy 47, 1554-1567. https://doi.org/10.1016/j.respol.2018.08.011

Silvast, A., 2017. Energy, economics, and performativity: Reviewing theoretical advances in social studies of markets and energy. Energy Research and Social Science 34, 4–12. https://doi.org/10.1016/j.erss.2017.05.005

Silvast, A., 2019. The role of social sciences and humanities in Norwegian environment-friendly energy research and innovation. In: Towards a global research agenda for transformative innovation policy conference, Valencia, Spain, 4–5 November 2019. https://www.tipconsortium.net/poster/the-role-of-social-sciences-and-humanities-in-norwegian-environment-friendly-energy-research-and-innovation/

Silvast, A., Hänninen, H., Hyysalo, S., 2013. Energy in society: Energy systems and infrastructures in society. Science and Technology Studies 26, 1–13. https://doi.org/10.23987/sts.55285

Silvast, A., Virtanen, M.J., 2019. An assemblage of framings and tamings: Multi-sited analysis of infrastructures as a methodology. Journal of Cultural Economy 12, 461–477. https://doi.org/10.1080/17530350.2019.1646156

Silvast, A., Williams, R., Hyysalo, S., Rommetveit, K., Raab, C., 2018. Who "uses" smart grids? The evolving nature of user representations in layered infrastructures. Sustainability 10, 3738. https://doi.org/10.3390/su10103738

Star, S.L., Griesemer, J.R., 1989. Institutional ecology, 'translations' and boundary objects: Amateurs and professionals in Berkeley's Museum of Vertebrate Zoology, 1907–1939. Social Studies of Science 19, 387–420. https://doi.org/10.1177/030631289019003001

TIP Consortium, 2017. Towards transformative innovation policy for Norway [WWW Document]. URL http://www.tipconsortium.net/transformative-innovation-policy-for-norway/ (accessed 5.24.21).

Winskel, M., 2018. The pursuit of interdisciplinary whole systems energy research: Insights from the UK Energy Research Centre. Energy Research and Social Science 37, 74-84. https://doi.org/10.1016/j.erss.2017.09.012

Winskel, M., Ketsopoulou, Irina, Churchhouse, T., 2015. UKERC interdisciplinary review. UKERC, London.

Wolsink, M., 2018. Social acceptance revisited: Gaps, questionable trends, and an auspicious perspective. Energy Research and Social Science 46, 287-295. https://doi.org/10.1016/j.erss.2018.07.034

CHAPTER 4

Calculating the 'Price' of Infrastructure Reliability in Finland

Abstract This final empirical chapter demonstrates how our Science and Technology Studies–inspired line of enquiry is also of use for considering the processes underlying and subsequent outcomes of large energy research projects, which have more conventional, monodisciplinary ambitions, and methodological tools, in comparison to the intentionally interdisciplinary projects discussed in Chaps. 2 and 3. Specifically, in this chapter, we explore a Finnish research project that aimed to study how much reliable electricity supply is 'worth' to the energy end-users, by assigning this reliability a financial price. Through discussing the experiences and outcomes of this project, we make clear how this reliability 'price' was translated and moved between survey studies, statistical modelling, and the needs of the energy industries and market regulatory profession. We conclude with direct discussion of how this chapter connects to the wider, interdisciplinary issues pertinent to this book, including boundary objects, the impacts of funding, epistemic cultures, and the importance of disciplines, and the implications of these for improving the understanding of technical and economic research projects that sit between vital public problems.

Keywords Market regulation • Blackouts • Government policy • Monodisciplinary energy projects • Economics • Power Systems Engineering

© The Author(s) 2022
A. Silvast, C. Foulds, *Sociology of Interdisciplinarity*,
https://doi.org/10.1007/978-3-030-88455-0_4

71

4.1   Introduction

This final empirical chapter provides an intentionally different perspective. Specifically, we leap into an empirical study that is markedly different from Chap. 2 (on the funder-driven whole systems agenda) and Chap. 3 (on the establishment of long-term, large-scale, multi-stakeholder energy research centres). Herein, we investigate how many of the questions and issues that were generated by these two previous chapters—around high-level commitments/investments to particular visions of *interdisciplinary* working—can also point us to better understandings of more conventional projects that, whilst addressing inherently multifaceted problems, still use *monodisciplinary* research tools. Indeed, we strongly assert that there is much to be learnt around these issues, including the different implied epistemic cultures within one discipline (in this case Power Systems Engineering), the impacts of funding to applied projects that serve regulatory policy-making and are hence cross-professional, the ways in which specific boundary objects (in this case, 'prices') mediate social worlds in research project settings, and ultimately how all such issues may shape or be shaped by surrounding disciplines and epistemics, even when not consciously considered within a relatively rigid monodisciplinary stance. This all matters, not least because the calls for and touted merits of interdisciplinarity are inevitably relative to its dominant monodisciplinarity cousin.

This chapter investigates such issues by exploring the outputs and experiences of a large, monodisciplinary, Finnish energy research project that was set up in the mid-2000s. The project sought out to study how much reliable electricity supply is 'worth' to the energy end-users, by assigning this reliability a financial price. The scholars, mainly from the Engineering Sciences, were given this task by the public authorities (then Finnish Ministry of Trade and Industry) and power companies that wanted to use the 'price' in the new regulation model of electricity utilities in Finland. The project was called KAH, short for "Keskeytyksestä aiheutuva haitta" and Finnish for "harm caused by an interruption (of electricity supply)". We henceforth refer to the project by its Finnish acronym, KAH.

The materials in this chapter stem from several sources, which indicate a double role as both project participant and its analyst—this is typical for Social Scientists in interdisciplinary projects, as we have outlined before in this book. First, one of the authors worked as a Research Assistant in KAH, gathering its data, conducting its surveys electronically and by phone, presenting the results to the funding body, and drafting the final report. Many of these reports and sources relevant to the KAH projects are also our materials in this chapter. Second, the author has examined this same project in a much larger study (Silvast 2017), whose related subsets have been published earlier (Silvast and Virtanen 2019). This larger study involved some 30 interviews with Finnish electricity professionals and lay-persons, and participant observation (Silvast 2018). Herein, though, we take this study in a new direction by interrogating the calculations of 'price' for infrastructure reliability as disciplinarily distinct research on a public energy policy problem. However, for the purposes of this chapter, only selected parts of these larger datasets are used. We use these interview and personal experience data to enable a richer, deeper consideration of the dynamics in play.

The structure of this chapter is as follows: we begin by presenting background context with regard to the economic-theoretical rationale for market regulation, which provides the foundations for the Finish project (Sect. 4.2). Further context is then offered on the importance of minimising blackout disruptions within the Finnish energy policy landscape (Sect. 4.3). Following this, we discuss the KAH project in more detail, with particular attention given to its production and use of cost estimates, in the context of, for example, objectivities, power dynamics, science-policy translations, and interdisciplinarity roadblocks (Sect. 4.4). We conclude by directly discussing this chapter's relevance for the pertinent issues being collectively generated by the three core empirical chapters (Chaps. 2, 3, and 4)—this cumulative progression in argument will, we hope, act as an appropriate stepping stone on route to Chap. 5, where we conclude the book with our presentation of a *Sociology of Interdisciplinarity* (Sect. 4.5).

4.2   What Is Market Regulation and How Does It Address Public Interest?

To understand the project context that sets the stage for this chapter, we must first visit the economic theory about energy networks and its market regulation in particular. Market regulation can be generally understood as a "negative feedback system in which asymmetries, whether due to social, economic or technological factors are balanced by rules which may (or may not) be codified in legal statutes" (Boyd 2001, p. 4). Separate from ministries and legislators in many countries, market regulators have "quasi-legislative power by allowing them to make rules having the force of law"

(Hirsh 1999, p. 31). Hirsh (2004) explains how regulators emerged in the US electricity industry in the early twentieth century, first and foremost to guarantee that electricity networks have adequate financial integrity. However, they also therefore enacted the notion that these networks are natural monopolies (i.e. services for whom there are no competitors):

> [T]he laws required regulatory commissions to maintain the financial integrity of utilities so they could expand their networks. To accomplish this goal, regulators ensured that utilities earned enough money to be able to pay investors attractive dividends and bond yields. At the same time, regulators guaranteed that competitors would not infringe on the franchise areas of utilities, thus explicitly codifying the notion that power companies constituted natural monopolies. (Hirsh 2004, p. 113)

In practice, hence, regulators are meant to ensure that public utilities (e.g. electricity networks) provide desired quality to their customers and are financially able to deliver that quality, especially when they will have no competitors. To address this 'natural monopoly problem', regulators have different regulatory instruments that they can use. Indeed, one previous piece of work (Boyd 2001, pp. 62–65) summarises a whole range of such instruments, some of which include persuasion and appeals to the public through the media, public operation, and ownership (instead of managing utilities privately); franchise bidding (running utilities as protected franchise for a set time); and whole deregulation (based on an assumption that gains from regulation are small compared to the costs of doing it). This said, the most common regulatory instruments are those focused on profits and pricing: namely, *profit regulation*, where utilities are allowed a specific rate of return, and *price control regulation*, where the prices to utility customers are controlled by the regulator.

As these various mechanisms show, regulation concerns far more extensive problems than merely efficiency and asset values. Regulation itself opens a set of issues around what is essentially a social contract between publics and utilities: where the utilities provide services that the public needs and the regulatory framework allows those services to be accomplished financially (Boyd 2001, p. 64). Questions around how exactly to govern these public interests, as well as what the public may need, have shifted over time, thereby introducing (typically monodisciplinary-framed) problems that are apt for study by interdisciplinary energy Social Sciences and Humanities (SSH) scholars—nevertheless, this topic has remained almost entirely underexamined, as we now go onto to discuss in Sect. 4.3.

4.3   LONG BLACKOUTS AND THEIR REGULATORY IMPACTS IN FINLAND

In the previous section, we made clear how traditional economic theory emphasises how market regulation is meant to ensure a quality service provision. In building on this, for the rest of this chapter, we use the problem of electricity quality of supply as an exemplar of public interests in reliable infrastructures, and we focus our enquiry on Finland. This section provides the supporting policy context for Finland.

Finland is a northern European state that has been facing difficult and long electric power failures for the past several decades (Silvast 2017). These have generated an active political and public debate on protecting the electricity infrastructure and making it as riskless as possible for blackouts. We assert that, in practice, it has been the Finnish public market regulator (paired with research-based insight) that has sought to translate these public interests into the economic values that the utility companies can act upon.

Difficult electric power failures have been a long-standing policy issue in Finland. Indeed, we can point to five past examples where they arose in public discourse and led to specific calls for market interventions. First, back in 2001, two exceptionally strong storms—Pyry and Janika—struck Finland and doubled the yearly number of electricity supply interruptions compared to the whole previous decade (Kauppa-ja teollisuusministeriö 2006, p. 29). In 2002, a publicly commissioned report on these blackouts (Forstén 2002) recommended that Finnish energy end-users receive compensations from all blackouts that last longer than 12 hours, and this became operational in the Electricity Market Act in 2003. The aim of this compensation entitlement was to "motivate electricity distribution owners to act in a manner that shortens the duration of interruptions" (Forstén 2002, pp. 31–32), and to this end, the report also suggested a maximum duration of six hours for an electricity interruption "even in exceptional conditions" (Forstén 2002, p. 2).

Second, four years later, the Finnish Ministry of Trade and Industry suggested that customers, or "entrepreneurs" critically dependent on electricity, purchase their own private emergency power generators (Kauppa-ja teollisuusministeriö 2006, p. 56), although a similar idea had been afloat since 2002:

> Uninterruptable electricity distribution cannot be guaranteed. If the customer's production or other activities do not tolerate reasonable electricity distribution interruptions, then the customer should personally secure the electricity supply. (Forstén 2002, p. 35)

Third, as unattainable as risk-free electricity distribution may be, since 2008 Finnish electricity network companies became additionally penalised financially for each electricity blackout, according to market regulation (Energiamarkkinavirasto 2007). This penalisation was done by linking power supply failures to the allowed yearly profits of the utilities. This hence supposedly gave them a financial incentive to improve quality levels.

Fourth, a major Finnish storm on Boxing Day 2011 initiated a blackout that momentarily affected 570,000 customers and lasted for days for tens of thousands of customers (Energiateollisuus 2012). Immediately afterwards, the power failure led to an untypically wide public debate concerning the crisis communication and crisis preparedness among private energy companies, the impacts of the outsourcing of their maintenance, and the necessity of preventing similar storm damages in the future by burying electrical cables—thereby considerably increasing monetary compensations for customer damages from blackouts.

Fifth, a new Electricity Network Law was enacted in 2013 (Electricity Market Act [9.8.2013/588]), stipulating that electricity network companies must make preparedness plans and set maximum durations for electric power failures. In order to meet these requirements, the network companies would have to invest over three billion euros into their networks (according to the Finnish national broadcasting company Yle 2020), most often by burying their distribution cables underground.

In all five of these examples, the public common effect of a blackout is transformed into a calculable risk in order to create a fair, transparent, and market-based way of distributing harms across all energy consumers.

The Energy Authority that regulates Finnish electricity companies rose to these challenges from a particular viewpoint. It was concerned with how all electricity network companies in Finland, especially the comparatively smaller companies operating in rural areas, could produce the necessary investments. Hence, its new regulatory model (from 2016 onwards) raised the allowed profits of electricity network companies. The result—scrutinised in a study by the Finnish national broadcasting company (Yle 2020)—was an increase in electricity distribution tariffs among many electricity end-customers; including those served by urban electricity companies, where power failures were less common and investment needs smaller than for rural areas. These higher tariffs were the subject of a public debate for several months.

The Energy Authority has defended the new model by stating that the rules should be equal to all network companies in Finland, especially since the model may have been challenged in market courts had this not been the case (Yle 2020). Thus, what is at stake is not only the regulatory formula—which has remained almost entirely opaque in the public debate—but also overarching questions concerning how to incentivise infrastructure providers to ensure service reliability, how to deliver fair profitability among natural monopolies, and how much electricity customers are willing to pay for improved reliability of infrastructures that they critically depend upon.

Yet to paraphrase questions raised several times in this book before: how can the regulatory experts know what the public needs and what their energy demand is like, or which parts of their activities most critically depend on functioning infrastructure? This is clearly a research problem where the SSH disciplines should have much to offer. A revealing quotation was given by Professor of Electric Power Technology, Pertti Järventausta, who was interviewed for the Yle (2020) report. Järventausta argued that the market regulator had concentrated on the network companies but had not taken a whole systems perspective that would have included impacts on the customers:

> The authority viewed this issue strongly from the perspective of network companies, so that they one would enable network investments and fulfill security of supply requirements. But now one forgot to conduct a holistic examination, what will this lead to in the customer end? (Järventausta, quoted in Yle 2020, no pagination)

This observation was clearly based on commissioned research Järventausta had co-conducted, where the first author was involved as a Research Assistant. This research study set to find out what indeed the customers need in terms of reliability, rather than taken that as a given.

4.4   Pricing Reliability in the Regulatory Model

This section briefly presents details of the original research study, KAH, which used monodisciplinary tools to examine what is a wide public matter: how members of the public value the importance of a critical infrastructure and how those valuations can be turned into prices that can be subsequently used in regulatory models. While such issues are often reported in project reports and publications, we offer unique access to them by drawing on the first author's own experiences in the project.

4.4.1   Connecting the Original Research Study to the Themes of This Book

Between 2004 and 2005, the first author was the Research Assistant for a large-scale research study that set to cover how laypersons perceive electricity blackouts (Silvast et al. 2006). It was commissioned by the Finnish Ministry of Trade and Industry and several power companies operating in Finland.[1] It was conducted by Electrical Technology Departments of two technical research universities: the Helsinki University of Technology (today merged with Aalto University) and the Tampere University of Technology (today merged with Tampere University).

The way to address power failure risk is commonly called Value of Lost Load, which is essentially a monetary estimate of the damage caused by power interruptions. In Finland, the monitoring of this damage in turn depends on information from energy users to uncover the measured economic worth of reliable energy supply. Such values are assessed in surveys like the one in Fig. 4.1, which we conducted in the study. Filled with questions about multiple blackouts and their economic effects, the survey assumes that all energy users are rational economic actors that calculate the value of energy use and the financial risk of electricity blackouts. The survey received some 1500 responses and its outcome was a complex set of averages of blackout values, across different kinds of customers (including households, agriculture, public sector, and industries, and including summer cottages as a typical Scandinavian category).

This survey (Fig. 4.1), as such, was not an exercise in interdisciplinary energy research. For example, it draws upon notions of rational

---

[1] Including E.ON, Fortum, Helsingin Energia, Imatran Seudun Sähkö, Kainuun Energia, Sata-Pirkka, Suur-Savon Sähkö, Turku Energia, and Vantaan Energia.

Question 16. Evaluate the economic value of the damages or harms that are caused to you by expected and unexpected power cuts. The electricity cuts are 1 second, 2 minutes, 1 hour, 12 hours or 36 hours long and they occur in winter during the week at the most harmful time.

power cut duration	unexpected power cut		pre-announced		the most harmful time (e.g. 18:00-20:00)
1 second	_____	euros			
2 minutes	_____	euros			_____
1 hour	_____	euros	_____	euros	_____
12 hours	_____	euros	_____	euros	_____
36 hours	_____	euros			_____

Fig. 4.1   A Finnish customer survey, 2004, asks what power cuts cost. More than a dozen similar questions are given in the survey. (Source: Silvast et al. [2006, p. 104])

consumers—common also today, for instance, in aspirations to make the electricity consumption more flexible by introducing dynamic time-of-use tariffs—that have been shown to be widely inadequate representations in social scientific research (Christensen et al. 2020). Further, this simplified survey did not always yield useful results, which may have been explained by the embedding of energy in everyday life. The household consumers, in particular, were not always able to estimate the monetary harms, especially in terms of short-lived interruptions. A report on the survey findings even notes this issue and, rather than including diverse viewpoints, ends up justifying the removal of 'outliers' that could not be explained by statistical averages:

> 79 % of respondents estimates the damage of one-hour unexpected power cut to be zero or did not respond at all. At the same time, 10% of responses in the same part was more than 100 euros. The largest response to this question as 1,600 euros … To find representative averages, the material had to be trimmed. We removed 10% of the biggest and the smallest responses, so that the average would represent the majority's responses. (Silvast et al. 2006, p. 47)

During project discussions, these outlier responses were commonly referred to as "subjective", as opposed to the seemingly "objective" answers that could be given by businesses, agriculture, and the industries on what damage power interruptions would cause. In fact, in the research group's scientific reporting on the study, the "subjective" answers remained mostly unaddressed. The publications and conference papers (Kivikko et al. 2007, 2008) turned to detailed statistical analyses and reporting of averages. An example is from the study results for the residential sector (Kivikko et al. 2007, p. 3): an unexpected electric power failure of 1 second would 'cost' 0.23 €/kW, for 2 minutes 0.84 €/kW, 1 hour 5.8 €/kW, and so forth up to 36 hours (costing 147.60 €/kW). The conference paper that includes these figures does not comment on the numbers and how they were formed in any manner, other than one noteworthy observation: that as can be seen, the householders' Willingness to Pay (WTP, i.e. how much more they would pay for reliable electricity at 1.10 €/kW for an hour's blackout) is only a fraction of the Willingness to Accept (WTA, i.e. how much more risks they would accept for a cheaper tariff at 8.30 €/kW for a similar blackout). In simple terms: if householders were perfectly rational economic agents as envisioned by economics, their WTP and WTA should be identical or very close to one another. The paper does not take on this difference and discuss it in more detail, although it does seem to indicate that people expect reliability to be higher than the actual costs of electricity that they pay for.

The first author did conduct some research on his own, interviewing and surveying householders (reported in Silvast 2017), which showed there were various kinds of blackouts and different people had a variety of responses to them. The acceptance of a blackout varied according to gender, to age, to region, and especially to memory about past blackouts. To be acceptable, a blackout also had to feel 'voluntary', rather than imposed from above. Such an acceptable electricity supply interruption, even though anticipated, should not halt those household practices that were perceived as important—and it also did not prevent less significant practices regularly or permanently.

In fact, temporality explained the seriousness of the blackout in at least three senses. First, a blackout should not interrupt everyday routines on a regular basis. Second, a blackout should not occur at a time when people have planned to do something else that requires functioning electricity. Finally, a blackout should not impact on tangible objects which are the result of time and investment (e.g. contents of a freezer and computer's hard disk drive).

Nonetheless, it was not possible to include these kinds of qualitative accounts as part of the scientific content of the report. In an interesting indication of interdisciplinary working and hierarchies of disciplines, there was an allowance to categorise the open qualitative responses that were given at the end of the survey study. However, even then, these categorised findings only became an Appendix (Silvast et al. 2006) in the final output, where they received only one table on one page of the 175-page report. The difficulty of qualifying the quantities of power cut damage persists in this research problem and also so in the regulatory domain. Undoubtedly, the monodisciplinary framing of this study held itself stubbornly strong throughout the project journey, including during the writing-up process discussed here.

4.4.2   How Did the Regulator Use the Cost Estimates?

The Finnish KAH study from 2006 (Silvast et al. 2006) was commissioned amid a change in the market regulatory model, which concerned not only Finland but European energy regulators at large. In digging deeper into the relevance and implications of this change for the Finnish economic regulation study that we have been discussing thus far in this chapter, we now draw upon CEER's (The Council of European Energy Regulators') published overviews of the European regulatory practices. CEER is a co-operation body of European national electricity and gas regulators.

According to CEER, many European countries' electricity regulation shared a common starting point until around 2000. This is because regulators operated through assigning price caps for the electricity network service that is billed from customers (CEER 2005, p. 31). Soon, however, the regulators noted that while managing one risk (overpricing), this mechanism created another risk. Specifically, even if prices are capped, electricity network companies might reduce their maintenance and investments to make a profit. And according to a popular line of thinking by economists (Gramlich 1994), lack of investment directly influences the quality of infrastructure provision: "Price-cap regulation without any quality standards or incentive/penalty regimes for quality may provide unintended and misleading incentives to reduce quality levels" (CEER 2005, p. 31).

New electricity regulation models, which are increasingly popular in Europe since 2005, strive to monitor and motivate improvements in this quality (CEER 2005, pp. 31–32). In practice, the theory would say that this means: statistics of quality are made public; "incentive" and "penalty" schemes are enforced so that utility companies control their profits in terms of their quality of supply; and a growing number of arrangements emerge that fix maximum durations for electricity blackouts and customer compensations for cases when the durations are not met. Along with compensations, however, the matter has also been about making customers aware of the costs of quality. Thus, specific emphasis has been paid to electricity customers' "expectations" and "their willingness to pay" for good-quality electricity (CEER 2011, p. 4). As has been summarised previously, "[r]esults from cost-estimation studies on customer costs due to electricity interruptions are of key importance in order to be able to set proper incentives for continuity of supply" (CEER 2010, p. 9). This is exactly the research problem that the original commissioned study (Silvast et al. 2006) set to address.

We now move to an example of concrete regulatory formula, to demonstrate how these issues are turned to activities in the regulatory profession. Specifically, between 2008 and 2011, the Finnish energy network quality regulation depended to a large part (although other measures were also deployed) on a method called Data Envelopment Analysis (DEA).[2] Developed in the US and situated in a scientific tradition called operational research, the method calculates the technical efficiency of multi-output, multi-input production units or decision-making units (Charnes et al. 1978). In so doing, it compares different decision-making units with one another, identifying the most efficient unit relatively and, in most contemporary applications, prescribing how the other units may improve their efficiency by altering their input, output, or both. While these aims may sound "neoliberal", the method's original intention was different: the DEA was developed to study "public programs" and "decision making by not-for-profit entities rather than the more customary 'firms' and 'industries'", and it depended on data "not readily weighted by reference to (actual) market prices and/or other economic desiderata" (Charnes et al. 1978, p. 429). Undoubtedly, such traits also made the method appealing to measure public electricity utilities and their efficiency.

This Finnish electricity regulation DEA model has the following formula for technical DEA 'efficiency', by which we mean: a utility's yearly outputs divided by inputs (Energiamarkkinavirasto 2007, p. 53):

---

[2] This example on electricity quality regulation and blackouts concerns Finland, but corresponding instruments were also in place in the mid-2000s in many other countries across Europe, including Sweden and Norway as well as the UK, Ireland, Italy, Portugal, Hungary, and Estonia (CEER 2005, p. 37).

$$\text{Data Envelopment Analysis}\,(DEA) = \frac{u_1 * \text{Energy} + u_2 * \text{Network} + u_3 * \text{Customers}}{v_1 * (\text{OPEX} + TP + KAH)}$$

Most of the inputs at this bottom of equation and the outputs at top of this equation are relatively common sense. Factors like a utility's operational expenses (OPEX, in the above formula) and property value depreciation (TP) are obviously a 'cost' from the quality point-of-view. A utility's 'productions' include the financial value of distributed electricity during a year (Energy), as well as the length of the utility's electricity network (Network) and the number of customers served by the utility (Customers) to normalise the utility's size. The variables—u_1, u_2, u_3, and v_1—are altered during a linear optimisation to maximise 'efficiency' relative to other utilities.

Along with these parameters, however, the customer's costs from blackouts (KAH) are an input. What does this mean in practice? Such costs have been first gathered by means of the surveys described in the previous subsection (Fig. 4.1). Based on these, the Finnish regulation model then concluded on the 'pricing' for blackouts (Table 4.1). For example, an unexpected electricity blackout would cost €1.10/kW of lost customer electric power and €11.00/kWh of lost customer electric energy. Other costs were assigned to planned interruptions and reclosing operations used by utilities to protect their systems that cause short-lived blackouts.

It is worth noting that the figures are not using entirely the same units and are not in the magnitude of the figures reported by Kivikko et al. (2007). There was a process of translation between the scientific work and regulation, where the regulator needed figures for whole of Finland and

Table 4.1 The regulatory 'pricing' of electricity blackouts in Finland between 2008 and 2011

Price/euros	Per kilowatt (power)	Per kilowatt hour (energy)
Unexpected interruptions	1.10	11.00
Planned interruptions	0.50	6.80
Fast reclosing operations	0.55	–
Delayed reclosing operations	1.10	–

Source: Energiamarkkinavirasto (2007, p. 34)

did not disaggregate them to different kinds of customers. It is very diffi-cult to find out how one cost became the other, and we assume that com-plex negotiations among electricity stakeholders took place although have no direct evidence of them.

Nevertheless, in the DEA input-output framework, such partly-researched, partly-constructed blackout 'costs' are then combined with managing electricity risk: the more costly the blackouts the customers have had, the more electricity the utility now has to distribute, or the less expense and property it has to have in order to appear 'efficient'. What emerges is a loop between customers' risk perceptions and potential for profit. This loop is furthermore performative: between 2008 and 2011, the Finnish Energy Market Authority set each electric utility an efficiency target to a large part based on a DEA formula (Energiamarkkinavirasto 2007, p. 49).

The "harm" (as per the project team) of a blackout and the techniques of its measurement have, we discovered, their own history. Often the "harm" it referred to was as the value of non-delivered electricity, not customer interruption harm like in Finland. According to an infrastruc-ture and electricity expert, who was familiar with the first Finnish studies that concerned these harms decades ago:

> the term interruption harm indicates that the customers experience blackouts as a harm and they should assess it. The perspective has not been similar elsewhere in the world, and one still hears talk about NDE [Non-Delivered Electricity] or such. Previously, in Finland, such NDE values were calculated without ask-ing customers. (Man, 60–69 years, National authority)

But the assumption in the DEA model is the opposite to NDE: cus-tomers are asked to calculate the level of risk and these are factored in as an input variable. All answers, or their averages, play a part in minimising electricity risk and distributing harms.

4.5   CONCLUSIONS

In this chapter, we examined the workings of translating values of public interest in electricity distribution between scientific research and market regulation. In doing this, we drew on a past Finnish energy research proj-ect—KAH, short for "Keskeytyksestä aiheutuva haitta" and Finnish for "harm caused by an interruption (of electricity supply)"—which was led by Power Systems Engineers who held an interest in economics (but were definitely not economists).

This chapter was intentionally different from the ones that preceded it, given that the KAH project was not interdisciplinary in its remit. To repeat, though, we would emphasise that monodisciplinary projects are worthy of exploration from an interdisciplinarity perspective. This is primarily because such projects do not exist in isolation: other disciplines, methods, and academic communities connect, especially when wide matters of public importance (e.g. functioning of energy infrastructure) are at stake. Acknowledging this, immediately allows for relevant cross-disciplinary questions to be posed, even to conventional and (some might say) narrow monodisciplinary approaches.

We want to now tie in this purposively different case to themes that have been emerging from this book thus far. First, we have demonstrated the obvious importance of disciplines. Indeed, the whole KAH exercise of valuing public interest in infrastructure reliability was seemingly conducted within one discipline: that of Power Systems Engineering. Yet, whilst it did not mention other disciplines by name, it was clear that during its analytical phase some type of applied economics (i.e. the study of prices and costs) had assumed the place of critical-SSH. That is to say, the notion of there being a 'cost' for reliability was a proxy for the public importance of functioning infrastructure, and this cost was not merely postulated in theory but became the topic of KAH's detailed empirical enquiry. This finding is not unique to the relatively esoteric topic of costs of reliability; it also appears, for example, when the national potential for energy efficiency and energy saving is translated to energy intensity (i.e. the ratio between energy output and Gross National Product). In all of these cases, many details (e.g. everyday practices of energy use and social norms) could not be acknowledged by the disciplinary scientific tools being used. We argue that it very much matters that economics was parachuted in to cover the societal elements of KAH, without due thought for its implications—as is emphasised by the common belief of SSH researchers that economics is ill-equipped to conceptualise and investigate matters of social order, and thus is why economics is fundamentally regarded as not being an SSH discipline (Foulds et al. 2017).

Second, however, while only one discipline worked on this topic, it did so by using different research tools and approaches, which were not always compatible. That is to say, we suggest different 'epistemic cultures' were at play, although this finding is not based on conventional ethnography

(Knorr Cetina 1999). As has been seen, the scientific version of statistics in the universities was different from the more applied economic regulatory models that the regulator wanted to create. For instance, the DEA regulatory model mentioned was driven forward by very different requirements (e.g. aggregated prices at the national level), whereas the Engineers in the project wanted to subject prices to detailed empirical enquiries. Even within KAH, they did not seem to agree on the level of detail and statistical sophistication necessary, with viewpoints diverging from the more pragmatic (what was needed to, for instance, complete the research) to the more explicitly scientific (what was needed to, for instance, publish in conference papers).

Third, the disciplinary concerns of the SSH—such as consumer research of electricity use—were wholly taken on board by a statistical style of reasoning, suggesting the dynamics of 'appropriation' (Forsythe 1999). Applied statistics also ended up essentially eliminating the putative 'subjective' answers that did not fit into the model of the rational consumer. This finding may relate to paradigms in scientific research; for example, the paradigm of Power Engineering cannot account for subjectivity; therefore, the more subjective observations can only appear as anomalies to the scientific method. This is interesting insofar it was not permitted to use these subjective answers within the KAH project, other than to a comment on methodological weaknesses, and there was no person in the KAH project dealing with the issue (including from a conventional economics, let alone SSH, perspective). Whilst it is true that the first author was allowed to work on the topic, this seemed to be under the normal arrangements of project work: the instrumental project work was prioritised and had to be completed first, and it was clearly the case that the study of 'subjective' answers was something he could write an academic thesis on (he was an undergraduate at the time). Even this arrangement, though, was not wholly clear-cut, as the author did, for example, present subjective findings in project board meetings and was invited to give workshops also in other contexts. The intriguing finding is that whilst SSH work was at one time seen as important, it was still nevertheless underprioritised.

Fourth, the cost features as a 'boundary object' (Star and Griesemer 1989): it is what mediates between the social worlds of public interest, market regulation models, and scientific research studies. It is also what is used to 'scope' out everyday life experience. As is often the case, and almost expected, the readers in key policymaking positions cannot account for the complexities of scientific methodological details, whether it be (in our speculation) due to, for example, lack of time, expertise, or simply an inability to use the methodologies in what they do. It is also important to note that while the costs were a boundary object for the public interest, the public was not allowed to engage in any manner in the issue within the project. The public interest was represented only by the constructed reality of economic models, and by offering economic replies that were quite trivially insensitive to everyday energy demands (i.e. the vital infrastructure of energy and social practices it sustains was visible only as abstract cost calculations). Thus, even while boundary objects do mediate between social worlds, they are a distorted proxy for them.

Fifth, we observed 'interpretative flexibility' (Pinch and Bijker 1984): the notion of 'cost' is itself quite flexible in how it can be interpreted. One important matter to notice is that while we have talked to 'cost' in this chapter, these were perhaps never meant to be 'actual costs'—although some research project participants probably still thought so. In contrast, a critical-SSH analyst would see the costs as constructed in the research process. For the power companies, the costs were possibly a proxy for their customer interests. For the regulator, we would argue that the costs were 'performative', in that they were meant to incentivise the power companies, not (necessarily) be based on real costs from actual consumers, whatever those may be.

Finally, we point to the key role of research funding. The KAH project was commissioned research by the Finnish Ministry of Trade and Industry and several Finnish electric power distribution companies. In the KAH project report (Silvast et al. 2006, p. 33), these were simply referred to as the "research funders". The funders did not generate an explicit interdisciplinary agenda, but it would be important to study what kind of an agenda they did create and how that might have impacted the project content. For instance, the KAH project had a steering board with representatives from the Ministry and the power companies that would actively steer the research work. We argue that the funders actively co-created the project, which is indeed more generally the case in applied technical settings, such as this. This situation furthermore made the inclusion of SSH highly difficult as a practical matter. Those engaging in interdisciplinary research should pay close attention to the restrictions imposed by (the required process of constructing) the project's design, as that can have tangible impacts on what kind of project work is considered to be relevant and possible, even if academically we know the issues demand interdisciplinary interrogation.

References

Boyd, K.J., 2001. Regulation and innovation: The case of metering in public utilities. Doctoral dissertation, University of Edinburgh. Edinburgh.

CEER (Council of European Energy Regulators), 2011. 5th CEER benchmarking report on quality of electricity supply. CEER, Brussels.

CEER (Council of European Energy Regulators), 2010. Guidelines of good practice on estimation of costs due to electricity interruptions and voltage disturbances. CEER, Brussels.

CEER (Council of European Energy Regulators), 2005. 3rd CEER benchmarking report on quality of electricity supply. CEER, Brussels.

Charnes, A., Cooper, W.W., Rhodes, E., 1978. Measuring the efficiency of decision making units. European Journal of Operational Research 2, 429-444. https://doi.org/10.1016/0377-2217(78)90138-8

Christensen, T.H., Friis, F., Bettin, S., Throndsen, W., Ornetzeder, M., Skjølsvold, T.M., Ryghaug, M., 2020. The role of competences, engagement, and devices in configuring the impact of prices in energy demand response: Findings from three smart energy pilots with households. Energy Policy 137, 111142. https://doi.org/10.1016/j.enpol.2019.111142

Energiamarkkinavirasto, 2007. Sähkön jakeluverkonhaltijoiden hinnoittelun kohtuullisuuden arvioinnin suuntaviivat vuosille 2008–2011 [Guidelines for evaluating the reasonableness of pricing of electricity distribution network owners for 2008–2011]. Helsinki.

Energiateollisuus, 2012. Loppuvuoden sähkökatkoista kärsi 570 000 asiakasta [570,000 customers suffered from the year's end's blackouts]. Press release 19 January 2012. Helsinki.

Forstén, J., 2002. Sähkön toimitusvarmuuden parantaminen [Improving the reliability of electricity distribution]. Helsinki.

Forsythe, D.E., 1999. "It's just a matter of common sense": Ethnography as invisible work. Computer Supported Cooperative Work 8, 127–145. https://doi.org/10.1023/A:1008692231284

Foulds, C., Robison, R., Balint, L. and Sonetti, G., 2017. Headline reflections—SHAPE ENERGY call for evidence. SHAPE ENERGY, Cambridge.

Gramlich, E.M., 1994. Infrastructure investment: A review essay. Journal of Economic Literature 32, 1176-1196.

Hirsh, R., 1999. Power Loss: The origins of deregulation and restructuring in the American electric utility system. MIT Press, Cambridge, MA.

Hirsh, R.F., 2004. Power struggle: Changing momentum in the restructured American electric utility system. Annales historiques de l'électricité 2, 107-123. https://doi.org/10.3917/ahe.002.0107

Kauppa-ja teollisuusministeriö, 2006. Sähkönjakelun toimitusvarmuuden kehittäminen: Sähkön jakeluhäiriöiden ehkäisemistä ja jakelun toiminnallisia tavoitteita selvittäneen työryhmän raportti [Developing the supply security of electricity distribution: The report by the working group that explored the prevention of electricity supply failures and the practical targets for the supply]. Helsinki.

Kivikko, K., Järventausta, P., Mäkinen, A., Silvast, A., Heine, P., Lehtonen, M., 2007. Research and analysis method comparison in Finnish reliability worth study, in: Proceedings 19th International Conference on Electricity Distribution Vienna.

Kivikko, K., Mäkinen, A., Järventausta, P., Silvast, A., Heine, P., Lehtonen, M., 2008. Comparison of reliability worth analysis methods: Data analysis and elimination methods. IET Generation, Transmission and Distribution 2. https://doi.org/10.1049/iet-gtd:20060532

Knorr Cetina, K., 1999. Epistemic cultures. Harvard University Press, Cambridge, MA. https://doi.org/10.2307/j.ctvxw3q7f

Pinch, T.J., Bijker, W.E., 1984. The social construction of facts and artefacts: Or how the sociology of science and the sociology of technology might benefit each other. Social Studies of Science 14, 399–441. https://doi.org/10.1177/030631284014003004

Silvast, A., 2018. Co-constituting supply and demand: managing electricity in two neighbouring control rooms, in: Shove, E., Trentmann, F. (Eds.), Infrastructures in practice: The evolution of demand in networked societies. Routledge, London, pp. 171–183.

Silvast, A., 2017. Making electricity resilient: Risk and security in a liberalized infrastructure. Routledge, London.

Silvast, A., Lehtonen, M., Heine, P., Kivikko, K., Mäkinen, A., Järventausta, P., 2006. Keskeytyksestä aiheutuva haitta [The harm caused by an interruption]. Espoo.

Silvast, A., Virtanen, M.J., 2019. An assemblage of framings and tamings: Multisited analysis of infrastructures as a methodology. Journal of Cultural Economy 12, 461–477. https://doi.org/10.1080/17530350.2019.1646156

Star, S.L., Griesemer, J.R., 1989. Institutional Ecology, 'translations' and boundary objects: Amateurs and professionals in Berkeley's Museum of Vertebrate Zoology, 1907–1939. Social Studies of Science 19, 387–420. https://doi.org/10.1177/030631289019003001

Yle, 2020. Yle tutki: Sähkön siirtohintojen raju nousu johtui yllättävästä syystä—valvojan tekemä muutos koitui kalliiksi kuluttajille [Yle investigated: Tsharp rise in electricity distribution prices was due to a surprising reason—the change made by the regulator was costly for consumers] [WWW Document]. URL https://yle.fi/uutiset/3-11588082 (accessed 5.25.21).

CHAPTER 5

A Sociology of Interdisciplinarity

Abstract In building upon the cases presented in Chaps. 2, 3, and 4, we develop a *Sociology of Interdisciplinarity* that draws our empirical insights together with resources from Science and Technology Studies (STS), in addition to Sociology of Scientific Knowledge, Research Policy, Infrastructure Studies, Anthropology, and Philosophy of Science. The key novelty of this framework is using STS insights to unpick the dynamics and consequences of interdisciplinary science, which distinguishes us from decades of earlier interdisciplinarity studies and gaps in understanding. Moreover, we not only focus on individual scholars and their experiences but pay careful attention to the wider contexts of interdisciplinary research, such as the impacts of funding structures, different access to resources, and power relations. We are careful in our approach so that our units of analyses—which vary from research groups and projects to whole epistemic communities and research policies—are most appropriate for the problem definitions that we put forward. The framework rests on a set of six dimensions, which we discuss in relation to current debates in the literature and our empirical analyses.

Keywords Funding structures • Epistemic cultures • Boundary objects • Disciplinary appropriation • Interpretative flexibility • Interdisciplinary energy research

© The Author(s) 2022
A. Silvast, C. Foulds, *Sociology of Interdisciplinarity*,
https://doi.org/10.1007/978-3-030-88455-0_5

5.1   Introducing a Sociology of Interdisciplinarity

We have examined the dynamics of large-scale energy research projects in three different cases: one on holistic interdisciplinary systems thinking in UK energy research (Chap. 2); another on interdisciplinary environmental energy research in Norway (Chap. 3); and a third case of a more conventional, albeit cross-professional, monodisciplinary energy research project in Finland (Chap. 4). Through these dedicated chapters, we have made clear the cases have been fundamentally shaped by the traits of the energy systems and policies in their respective countries, as well as also consequently reflected upon the various different configurations of (interdisciplinary) research practice in play.

In this concluding chapter, we now draw our key lessons together through synthesising a new framework, a *Sociology of Interdisciplinarity*. Our intention in presenting this framework is to put the analytical spotlight firmly on the social dynamics of (doing) interdisciplinarity, in a bid to spark further inspiration for scholars and practitioners in their future work. We would also hope, despite much of this book being built on a bedrock of interdisciplinarity in energy research, that this framework is of direct use and interest to all those interested in well-functioning interdisciplinary research systems. This could include, but not be limited to, managers of funding programmes, research evaluators, administrators, policy officers, and the like.

Our core argument in this book is that interdisciplinary research should be studied as social activity and the scientific ideas that it generates explained by sociological dynamics. This general interest is not novel: it has been the ground for decades of Social Sciences and Humanities (SSH) research programmes. Indeed, an interest in 'interdisciplines', which integrate conventional academic disciplines, and 'problem-based science', which cross disciplinary boundaries, dates back several decades in research and higher education policies (e.g. Barry et al. 2008; Gibbons 2000; Nowotny et al. 2001; Klein 2010). These themes have remained important in current discussions about interdisciplinary collaborative teamwork (Balmer et al. 2015).

A large literature on this topic has generated important insights—such as the differences between *interdisciplinary, multidisciplinary*, and *transdisciplinary* research and their approaches and methods (Klein 2010). We visited these approaches when introducing our position on key debates in the literature, in Chap. 1 (Sect. 1.2.1), but it bears repeating that interdisciplinarity is a complex concept and does not have a single definition. Nevertheless, certain taxonomies pertain to it, for example, between theories and methods that are interdisciplinary (i.e. integrating different academic disciplines in knowledge production), multidisciplinary (i.e. juxtaposing disciplines but keeping their original identities), and transdisciplinary (i.e. transcending disciplinary-based knowledge altogether).

The discourse on interdisciplinarity and multidisciplinarity is not merely of theoretical, academic debate, but is also based around real ground-level experiences of doing funded research. Indeed, it forms an increasing part of how energy researchers now work. In 2009, for example, the UK's network of academics UK Energy Research Centre (UKERC) produced an overview of the interdisciplinary research centres in the country (Wang 2009). At this point, the number of new energy centres and networks that cut across standard university departmental and faculty structures was already increasing visibly. The report found nearly 40 cross-departmental networks, interdisciplinary centres, and cross-institutional collaborations across the country. When this document was updated in 2019 (Silvast 2020), the activity around interdisciplinary energy centres and networks in the UK had grown so large that it would no longer meaningfully fit into one review. A response in this updated review was therefore to no longer represent all interdisciplinary initiatives and networks, but instead pragmatically detail research projects that were performing interdisciplinary agendas.

Relatedly, Mark Winskel (2018) has brought together a conceptual interest in interdisciplinarity with the pursuit of energy research in the UK, through various activities under the UKERC banner. He highlights several main choices and trade-offs in interdisciplinary energy research, including disciplinary diversity, integrated knowledge production, and how much non-academics (e.g. industries, publics, and policymakers) participate in the co-design of research (which Winskel names as *transdisciplinarity*, rather than mere *interdisciplinarity*). Our book builds on similar insights considering how interdisciplinarity is being configured in particular energy projects. It builds from Science and Technology Studies (STS)—an underutilised perspective in the study of interdisciplinarity in energy research.

Working from these perspectives, this book has provided a detailed examination of how interdisciplinary energy research has been conceived, and what consequences and dynamics it has had especially to those involved in interdisciplinary research projects themselves. It produced fresh insights into the lived experiences and actual processes underlying interdisciplinarity, rather than how it is being merely explicitly advocated. To accomplish this goal, we presented empirical studies on large-scale energy research projects set between academia, public policymakers, and industries, using mixed SSH methods ranging from ethnographic field-work and qualitative interviews to desk-based research and literature reviews. These accounts recounted how interdisciplinarity works in prac-tice, from the perspective of those carrying it out—what works, what does not work, what are the challenges, and so on—which are increasingly rel-evant given the prevalence of and very real steer for most energy scholar-ship to be interdisciplinary. We provided ground-level experiences of how interdisciplinarity is done, from an empirical perspective: providing inter-esting stories and experiences that energy research(ers) can relate to.

Our particular aim has been to move between different scales of inter-disciplinarity and explain how these scales are interconnected: from the experiences of scholars, on the one hand, to the impacts of funding struc-tures, the epistemic cultures that produce knowledge on energy issues, and the social dynamics of research projects on the other hand. This book's key contribution is in designing and presenting a new framework, a *Sociology of Interdisciplinarity*, which combines our results and draws insights from various literatures to unpack interdisciplinary research in practice.

The framework rests on six dimensions (Table 5.1), each of which is discussed in turn in the following sections, in relation to the literature and this book's empirics: the impacts of funding (Sect. 5.2), epistemic cultures (Sect. 5.3), boundary objects (Sect. 5.4), appropriating disciplines (Sect. 5.5), interpretative flexibility (Sect. 5.6), and the importance of disciplines (Sect. 5.7).

Before an explanation of each dimension, we believe it useful and instructive to offer some supporting advice on our proposed use of this framework:

- The dimensions are not intended to be comprehensive, but instead represent some of the main issues that spoke to us through the stud-ies in this book. There are inevitably other dimensions, and we are eager to be pointed towards them. We have selected six sets of issues that we think do not get the attention that they deserve, despite STS literature indicating their potential fruitfulness.

Table 5.1 A Sociology of Interdisciplinarity and its six dimensions

Dimension	Explanation
1. The impacts of funding	Research funding has effects in bringing about certain kinds of working practices, research teams, and research outputs.
2. Epistemic cultures	Interdisciplinary projects produce knowledge in specific epistemic cultures—knowledge-oriented groups of scholars—that cut across broad academic disciplines (e.g. Engineering, Physics, various SSH).
3. Boundary objects	Knowledge moves between the epistemic cultures in interdisciplinary projects via special boundary objects (e.g. computer models, calculations of risk).
4. Appropriating disciplines	Interdisciplinary projects can involve the more powerful disciplines appropriating the tools and methods of other disciplines.
5. Interpretative flexibility	Interdisciplinary projects create a ground for more disputes about how 'facts' and technologies should be interpreted. While interdisciplinarity is often favoured by funding bodies and researchers as a label, this also conceals the considerable interpretative flexibility of the concept itself.
6. The importance of disciplines	Continued importance of conventional academic disciplines in interdisciplinary contexts.

- One is free to use this framework in any way that they see fit: we envision that uses range from zooming in on one particular dimension, to covering some or all of them in parallel to understand a specific research programme or project.
- The framework was clearly built for critical-SSH analyses, but we are fully committed to them enhancing reflexivity in practical decision-making situations. This could involve, for example, designing a new funding instrument, interdisciplinary evaluation, starting an interdisciplinary programme in a university, or many other uses.
- Lastly, we emphasise this as one Sociology among others—*a* Sociology of Interdisciplinarity and not *the* definitive Sociology. We very much hope that colleagues will offer critique on these dimensions, drill deeper through further empirical contexts, and offer evidence to these dimensions. What follows is therefore a starting point to stimulate reflections on working in an interdisciplinary manner.

5.2   The Impacts of Funding

The first dimension of our framework provides viewpoints for examining university research funding and how the source of funding affects scientific activities, in this case, interdisciplinary research projects. Previous scholarship addresses this issue as the structuring impact of funding (Salmenkaita and Salo 2002). The key arguments about the (un)intended consequences of competitive research funding (Geuna 2001) and the underlying funding negotiations processes (Davenport et al. 2003) are now several decades old. That increased reliance on external funding and competing for it affects research output is—or has now also become—common sense among academic scholars. Yet, empirical research has demonstrated that the outputs of researchers reliant on applied and externally-funded projects do shift in the longer term (Goldfarb 2008). Comparative studies show this especially between scholars mainly working in university-funded projects versus externally-funded projects. The general observation is that if projects require a large degree of support from non-academic sponsors and partners, this impacts upon the outputs that researchers prepare, especially journal articles (Manjarrés-Henríquez et al. 2008).

Such arguments certainly fit with well-established ideas in STS around how documents carry agency—in this case, how funding documents (call texts, proposal templates, etc.) are actively scripting responses from those applying for said funding, and thereby also directly shaping its subsequent outcomes and recommendations. As Royston and Foulds (2021, p. 3) put it, such "documents contextualise the goals that frame energy research, and simultaneously enact and embed—albeit through complex, political and negotiated processes—the knowledges produced by research". Indeed, this has been the rationale behind a number of recent studies that have called for greater diversity in energy research and innovation funding, with a particular emphasis on moving away from natural/technical science-based or techno-economic solutions (e.g. Foulds and Christensen 2016; Genus et al. 2021; Overland and Sovacool 2020).

In general, though, the points that past scientometric and research policy studies make, however, are still focused on disciplinary-based research systems (c.f. Winskel 2018), just with a marked and increasing reliance on competitive external funding—which many argue is only becoming more uncertain and fickle, as a result of its increasing normativity (c.f. Foulds et al. 2021). Furthermore, it is indeed true that many interdisciplinary energy projects are reliant on major and long-term external grants. These kinds of large grants correspond with a pertinent requirement of interdisciplinary projects, as recognised in the UK: the need to learn between disciplines and coordination/maintenance of interdisciplinary collaboration (Hargreaves and Burgess 2009; Longhurst and Chilvers 2012). But these kinds of grants also have specific values and priorities attached to them. During the past years, interdisciplinary grants have been especially linked to delivering fresh insights on grand societal challenges, notably energy and climate issues. Examples are easily available for all to see, such as the European Commission's mainstreaming commitment to SSH (Kania and Bucksch 2020), which has often led to the deployment of forms of SSH that are "minimal, disciplinarily-narrow, overly-instrumental and lacking [of] critical perspectives" (Foulds et al. 2020, p. 5) within large-scale interdisciplinary energy projects.

The Norwegian Centres for Environment-Friendly Energy Research (FMEs) (examined in Chap. 3) were evaluated by a strategic and finance consultancy (Impello 2018), highlighting important findings on the instrumental outcomes that they were meant to achieve. The main research question of the evaluation report was whether public research funding to energy research has been worth it in monetary terms. While it acknowledges research results (e.g. knowledge, concepts, and systems), the report's main conclusions are about impacts that are quantifiable: that is, calculations on the future impacts of energy research to emissions, energy use, economically, and for innovativeness. Here, large grants were expected to generate measurable effects to energy provision and the economy (while the report includes a category of other qualitative effects, which includes impacts on society, these are only assessed as binary—i.e. as detected or not detected).

The evaluation does not explicitly talk about the effects of interdisciplinary collaboration and explicitly leaves the social scientific FMEs outside of the evaluation (Impello 2018, p. 22). Our literature reviews in Chap. 3 point to the prevalence of social scientific and sociotechnical publications on FMEs appearing as grey literature (e.g. reports, evaluations, and student theses), taking precedence over peer-reviewed articles on this topic. Resembling the scientometric studies that observe support duties externally funded projects (Manjarrés-Henríquez et al. 2008), the interviews with FME members recognised Social Scientists as often becoming their administrative coordinators, communication experts, or supporters studying, for example, innovation or consumer acceptance. Complementing this focus, the Finnish case in Chap. 4 points to how scholars working for external funders became the producers of policy-relevant costs and figures, where the academic methodological considerations on these figures (and the social scientific qualitative valuations related to them) were often difficult to translate to direct policy and regulatory relevance. All of this suggests that the projects with large grants configure particular roles for SSH research—also recognised in the UK whole systems research (Mallaband et al. 2017).

Concurrently to large grants, however, some funding agencies have encouraged interdisciplinarity in almost diametrically the opposite way. Some of them engage in short-term, facilitated projects to encourage interdisciplinary research. An example from the UK is the Sandpit funding model, developed a decade ago by the Engineering and Physical Sciences Research Council (EPSRC) in the context of whole systems research. In the sandpits, which lasted only a few days, invited academics held workshops and engaged in brainstorming with the eventual aim of generating competitive bids for large-scale grants. This short-term model was said "to bring individual academics together who would not, under normal circumstances, be likely to meet and share ideas" (Hargreaves and Burgess 2009, p. 8). Here, the idea is that unconventional ideas will be encouraged by transgressing normal ways of working within academic disciplines.

This general model of short-term sandpits can also be observed within large granted research projects. One possible manifestation of it is the recurrent reliance on workshopping and project meetings to come to more relevant interdisciplinary themes for research. Notably, nearly all of the findings of interdisciplinary working in the CESI project in Chap. 2 were drawn from various kinds of project meetings: events where academics from different disciplines were required to come together and share their ideas on a set common theme, such as energy demand research, future energy scenarios, or policy relevance of energy modelling tools. The ideal seems to have been that interdisciplinary knowledge production happens in encounters during the meeting, and that one crucial output of it is an organised dialogue in itself. Workshop notes were collected and shared as one relevant material outcome of these encounters. Conversations and presentations held at workshops also sometimes became the resource for further project work, as earlier workshop conversations would affect future workshops or concrete decisions being made on the direction of the modelling tools.

Another, and a third, potential way that funding may affect interdisciplinary working is the need to frame disciplinary differences in ways that are recognised by grant funding agencies. Interdisciplinary research is, by definition, unconventional. However, as two scholars in the Social Studies of Science note, "to gain funding for such research, scientists are forced to outline unconventional ideas in ways that still relate to recognised concepts and findings, as well as adhering to the conventional requirements of relevant fields of research" (Philipps and Weißenborn 2019, p. 884). Indeed, from our own proposal-writing experiences, we can certainly point to numerous examples of colleagues creatively relabelling themselves along disciplinary lines, in a bid to align with funder expectations.

Furthermore, STS scholars have also collaborated with Julia Thompson Klein—the noted interdisciplinary studies scholar—to examine how interdisciplinary research was carried out in project proposals funded by the Academy of Finland (Huutoniemi et al. 2010). They noticed that ideal types and conceptual categories of interdisciplinarity were common, as was reliance on names of disciplines in grant applications. Instead of these explicit labels, the study scrutinised interdisciplinary research content—focusing on how research had crossed conventional bodies of knowledge, concepts, methods, and research practices. This study came to constructive conclusions on genuine interdisciplinary work rather than mere teams of interdisciplinary scholars. Interactions between research fields happened frequently and were substantial for the examined projects. But to discover this, the scholars had to look at research content itself, thereby going beyond how disciplinary and interdisciplinary labels were explicitly articulated.

This finding directly brings us to the next dimension and the need to study how interdisciplinary research is being carried out in project work, more than the labels given to it.

5.3   Epistemic Cultures

The concept of epistemic cultures was popularised by anthropologist Karin Knorr Cetina (1999), who developed the concept in her detailed ethnographic study of two fields of science: that of High-Energy Physics and Molecular Biology. Epistemic cultures can be defined as units that "produce and maintain specific understandings of what valid knowledge is and how it should be produced and understood" (Kruse 2021, p. 3). Knorr Cetina based her study of knowledge production on ethnography of scientific laboratories and their working practices and cultures, but the concept has been since applied to a variety of problems in STS, ranging from citizen science projects (Kasperowski and Hillman 2018) to forensic evidence

(Kruse 2021). The underpinning issue with such studies is that they have been developed through using epistemic cultures to investigate scientific disciplines, which aligns with the agendas of this book.

Knorr Cetina's (1999) approach to disciplines is constructive: disciplines are apt for addressing how science is organised. Yet, they offer less cogent descriptions of expert practice, which is why she coined these practices as 'epistemic cultures':

> In the past, terms such as discipline or scientific speciality seemed to capture the differentiation of knowledge. The notion of a discipline and its cognates are indeed important ones in spelling out the organising principles that assign science and technology to subunits and sub-subunits. But these concepts proved less felicitous in capturing the strategies and policies of knowing that are not codified in textbooks but inform expert practice. The differentiating terms we have used in the past were not designed to make visible the complex texture of knowledge as practiced in the deep social spaces of modern institutions. To bring out this texture, one needs to magnify the space of knowledge-in-action, rather than simply observe disciplines or specialties as organising structures. (Knorr Cetina 1999, pp. 2–3)

This stream of research on epistemic cultures connects well with our findings from our three cases on interdisciplinary energy research. Indeed, the interdisciplinary energy agendas that we have observed were not directed towards disciplinary (sub)units that are contained in textbooks and should still inform the strategic discourses of interdisciplinary researchers. Instead—as we explain further in Sect. 5.7—the agendas focused on how researchers work, using various methods and tools, reasoning, and other elements in their "machineries of knowledge construction" (Knorr Cetina 1999, p. 3). This focus includes how such knowledge practices are assigned with cultural significance in different contexts.

To be clear, Knorr Cetina's work on epistemic cultures examined the diversity of scientific laboratories and, in doing so, deep-dived into the detailed ways of working, attributing of scientific authorship, and collective structures in these normally restricted field sites. This book has also used ethnography to study the inner life of collaborative interdisciplinary teams—especially in Chap. 2—but applied it more strategically, complementing it with other data sources (e.g. written documents, researcher interviews, and literature reviews) especially in Chap. 3, and drew on data gathered whilst working in a conventional applied project in Chap. 4.

With these clarifications in view, we strongly advocate the concept of epistemic cultures as a means by which to explain the dynamics of interdisciplinary research in these three settings.

To us, epistemic cultures offer an appropriate description of the differences that arise when various teams come together to solve problems in large interdisciplinary energy projects. In Chap. 2, it helped point out that the science of energy modelling, now widely applied in interdisciplinary projects, is in itself disunified, divided into subcommunities of scholars that model a particular subsystem of the whole energy system, and attribute meanings differently to their findings. The point is not that they are different subunits of one discipline—such as subdisciplines of the Physical Sciences and Engineering—but that their knowledge about the energy system is made in a widely different manner, giving rise to different epistemic cultures (Silvast et al. 2020).

We saw these epistemic differences manifest between so-called energy demand modellers on the one hand, who are attuned to the intricacies of energy use in everyday life and the uncertainties of measuring it, and more conventional energy supply modellers on the other hand, for whom energy demands of everyday people appear as statistical properties to be fed into the energy computer models and solved as part of flow equations. We would argue that the salient difference was in their machineries of producing knowledge: in this case, what the model knew of the target system.

Chapter 4 showed similar epistemic differences among the scholars who all work in power systems technology. At least three epistemic cultures had influenced the practice of producing more scientific costs for the reliability of electricity infrastructure: one focused on statistical methods; another on economic modelling; and a third took an empirical, mainly pragmatic, approach to these costs. Each of these cultures found it challenging to work with another culture to solve problems, hence highlighting similar problems between knowledge production within one and the same discipline that would be normally addressed as problems in working between academic disciplines.

The differences observed in this book between Energy Social Scientists and Engineers, such as modellers, can be also explained by their different epistemic cultures: concerning not only the obviously different ways in which these scholars produce knowledge, but also differences in attributing authorship and working collectively (i.e. epistemic cultures). There is a distinct contrast between the more collectively focused, collaborative-modelling communities and the Social Scientists and Humanities scholars that often work in a solitary way or as part of small dynamic groups of colleagues (Beaulieu 2010). These differences manifest, for example, when both epistemic cultures are asked to come to solutions to future energy issues, and the steps taken may be entirely different across them.

Different epistemic cultures may also go some way to explain disciplinary balances in interdisciplinary energy projects. In Chap. 3, we saw that Social Scientists had constituted the minority in some of the early Norwegian Centres for Environment-Friendly Research (FMEs). While various reasons may be behind this resource distribution, one possible explanation can once again be offered by epistemic cultures. Namely, the culture of some Engineers may be based on the assumption of large teams needed to work collectively to run models and to experiment with technologies, whereas Social Scientists are expected to work in relative isolation. The solution to change these disciplinary balances within interdisciplinary projects is therefore not only a simple matter of redistributing resources, but it also requires acknowledging the different cultures of knowledge production, and how those differences can be understood and appreciated.

But if epistemic cultures successfully highlight the inner life of particular scientific knowledge production within a bounded working culture, there is one key limitation that pertains to the concept. This, as Kruse (2015, p. 110) remarks, is that the concept "does not address the question of how knowledge might travel between epistemic cultures". This critique is highly pertinent to our cases, as most interdisciplinary energy projects are exactly about the exchange of knowledge between different knowledge-oriented cultures, or whole disciplines. We need more and different conceptual resources from STS to more fully address this issue, which takes us to our next dimension: boundary objects.

5.4   Boundary Objects

In Chap. 2, we visited the field of UK energy modelling and learnt about some of the models used and being developed, especially by the EPSRC-funded Centre for Energy Systems Integration. Another known modelling example in the country, though now superseded, is the MARKAL,[1] which was also reviewed in the same chapter. The MARKAL model was advanced by academics but applied in energy and climate policy at large. It was also

[1] MARKAL: MARKet and ALlocation.

in use for a remarkably long period of time; starting from early developments in the 1970s, to it still being used in the late 2010s. UK government policies and priorities changed markedly during this time: from energy systems analysis and strategy development during the 1970s oil crises, for example, all the way to energy market liberalisation in the 1980s, and then current decarbonisation goals.

What made this energy computer model successful for so many stakeholders and in a rapidly changing policy environment? One possible explanation could be its sheer complexity and wide scope. The University College London (UCL) Energy Systems Team has claimed rigour and credibility because of the model boasting half a million data elements, and because of the wide-ranging extent of the energy system that it models, from production resources to fuel processing, infrastructures, conversion, end-use, and service demands (UCL 2021).

This complexity of the model is impressive in itself considering that all energy models are necessary simplifications of the system that they represent. Yet, and intriguingly, four UK scholars, one of them based at the UCL, have offered a somewhat different interpretation of the 35 years of history of the MARKAL model (Taylor et al. 2014). They argue that MARKAL was successful mainly because it functioned as a 'boundary object' (Star 2010). This means that MARKAL facilitated dialogue—in practice, bringing together communities of practice with various institutional and professional logics. The examples that the authors use are academics and policy practitioners. While arguably sharing the same overarching goal of decarbonisation, these communities can differ in respects to the rationales of how this goal should be reached: where academics want to introduce more debate, for example, policy practitioners may need to close the debate to make decisions. Nevertheless, as a boundary object, the model can fulfil both these goals at the same time as well as allowing knowledge integration across the boundary.

Boundary objects are artefacts, concepts, or methods that lie at the interface of different social worlds, such as politics and the economy. They are also potentially at the interface of different epistemic cultures that are special kinds of social worlds, which are made coherent by their members working with the same specialised tools and technologies (Clarke and Star 2007). Boundary objects facilitate co-operation and coordination between such social worlds because the identity of these objects—even if not all their details and intricate functioning—is understood across these worlds. In addition to computer models, earlier literature in energy research cites databases, standardised methods, and forms as examples of boundary objects (Taylor et al. 2014). The cases in this book offer several similar and other further examples.

For example, in Chap. 2, we presented computer models that are functioning, or are at least expected to function, in very similar manners to MARKAL in the UK. In fact, these models (as boundary objects to policy decision-making) received specific and explicit attention among the project members, even if not in the same terminology. The concept of boundary objects introduces a novel interpretation of how this policy relevance was happening. We could, rightly, critique the energy systems modellers for not developing a deeper understanding of how policy processes and governance work and the manifold cycles involved in them. Yet, we could also argue that the very point of models as boundary objects is that they act as intermediaries between these social worlds. Neither the policy community needs to understand the intricacies of academic energy modelling, nor can the modelling community understand the details of policy processes, but the boundary object fulfils both logics and rationales. In Chap. 2, the same could be said about the concept of energy demand and future energy scenarios: objects and methods whose identity was understood across different parts of a large project, and that actually facilitated the project's co-operation and coordination, even if there were many implicitly different interpretations of what these concepts could mean. That is to say, there are simplifying, reductive steps involved in interdisciplinary exchange that allow knowledge trading to happen.

In Chap. 4, the calculations of reliability costs for infrastructure, and the very concept of there being such costs, resemble boundary objects (see Silvast and Virtanen 2019). Market regulators interpret the costs to be about meeting the expectations of consumers with their 'willingness to pay' for electricity reliability. This cost is 'performative' as its main purpose is for the energy distributors to internalise the need to be reliable: it does not have to be an 'actual cost', although that is obviously of value. The researchers studying these costs interpret them to be an empirical phenomenon that exists among real consumers out there, or an object to be examined by statistical methodologies. The notion of there being costs for breakdowns of a ubiquitous critical infrastructure service is shared by all the involved professional communities, yet the rationales for using and examining these costs differ. Nevertheless, not one of them would claim that there are no reasonable costs to be found, although this would be a typical social scientific critique of the rational costs of energy use in everyday life (e.g. Christensen et al. 2020). In other words, a shared level of credibility is required for working across social worlds in projects.

In the Norwegian case in Chap. 3, the concept of environmental innovation was also similar to a boundary object, especially during the early days of the large centres studied. Even if innovation initially had no single interpretation, its identity was understood in a somewhat coordinated (yet flexible way) that could be yielded by the knowledge production tools of the respective epistemic cultures. In this setting, as shown by the evaluation (Impello 2018), innovations could indeed mean commercialised inventions as is traditionally the case (Schot and Steinmueller 2018), but they could also encompass new tools and methods such as energy systems analysis, optimisation and simulation, definition guidelines, and even handbooks for environmental design. We could critique the innovation term for lacking conceptual depth, but boundary objects take the argument in a different direction: namely that as boundary objects, these innovations were likely able to coordinate the group activities and generate coherence among them. As Taylor et al. (2014) showed with the MARKAL, boundary objects can function to connect professional communities even when they do not have a necessary disagreement, but simply have to operate in different institutional and professional contexts.

In the next section (Sect. 5.5), we study further the moving of methods and approaches between different social worlds and epistemic cultures, especially relating to Social Sciences and other disciplines. This qualifies how, in contrast to boundary objects that mediate between social worlds, translating scientific methods between the confines of disciplinary identities is not always frictionless and unproblematic. In building on this still further, Sect. 5.6 then discusses how scientific and technological disputes may relate to interdisciplinary projects.

5.5   APPROPRIATING OTHER DISCIPLINES

Diane Forsythe's famous research in STS was situated at the intersection of Medical Informatics, Computer Sciences, Ethnography, and Anthropology—hence why she operated at the interstices of the Natural Sciences and Social Sciences, leading to her research being fundamentally interdisciplinary by design. In describing these interdisciplinary collaborations, which tended to centre on user studies, she coined the term 'appropriation'.

In contrast to its current use in 'cultural appropriation', Social Scientists have traditionally used the term in a more neutral manner. It refers simply to people acquiring new things, such as consumer products. In this sense, for example, it can be said that fitting new Information and Communication Technologies to people's lives is 'appropriation', where the main reference is to the negotiations and considerations that led to acquiring these technologies (Haddon 2011). This more neutral meaning of the term also characterises how we use it in this book, since we want to avoid claiming that there are some actors appropriating the property of others in collaborative projects.

Forsythe (1999, 2002) made an intriguing discovery in her work with Medical Informatics, working in these fields herself in the role of an Anthropologist. She noted the increasing prevalence of anthropological ethnographic methods that supported software design since the 1970s. While this led to an increase of trained Anthropologists employed by research laboratories and companies, it also had another unforeseen consequence: non-Anthropologists, such as Physicians and Computer Scientists, started to borrow ethnographic techniques in their own work. That is to say, the dynamics of appropriation emerged. While such borrowing is not inherently problematic, she argues that ethnographic expertise was lost during translation, meaning that social scientific methods became misunderstood at large as a result. Building on experiences from such collaborations, Forsythe (1999, p. 130) summarised what she called "six misconceptions about the use of ethnography in design":

1. Anyone can do ethnography—it's just a matter of common sense.
2. Being insiders qualifies people to do ethnography in their own work setting.
3. Since ethnography does not involve preformulated study designs, it involves no systematic method at all—"anything goes" (p. 130).
4. Doing fieldwork is just chatting with people and reporting what they say.
5. To find out what people do, just ask them!
6. Behavioural and organisational patterns exist "out there" (p. 130) in the world; observational research is just a matter of looking and listening to detect these patterns.

Forsythe (1999) proceeds by correcting and qualifying these misconceptions, and we draw from some of that now, but the discussion on the 'proper' use of social scientific methods is not in the direct agenda of this book. What is more interesting here is asking why appropriation had happened in these collaborations, given that they had also typically perceived the Social Sciences as 'new' and 'soft' disciplines in the very same settings (Forsythe 2002). There were myriad reasons for the appropriation, some of them to do with intellectual curiosity and some with the increasing awareness about the Social Sciences as such. Firstly, Medical Informatics involved Physicians, Nurses, Medical Librarians, Computer Scientists, and Information Scientists, hence being at the interstices of Medicine and Computer Science. These were, for a key part, influential and highly educated experts, themselves knowledge workers, some of them with doctorates both in Computer Science and in Medical Science, and already working in interdisciplinary manners. Secondly, many of them had been routinely exposed to the Social Sciences: namely, having had "some acquaintance with ethnography from reading publications that draw on ethnographic research, hearing talks at professional meetings, working with social scientists on research teams, and/or being subjects of ethnographic inquiry themselves" (Forsythe 2002, p. 145).

In other words, more generally, it is the exposure to social scientific ways of working that enabled the influential and interested knowledge workers from other disciplines to appropriate the social scientific method to their own work. This explanation also clearly seems relevant to the interdisciplinary energy research collaborations we have been studying in this book. It was especially pertinent to Chap. 2's CESI 'demand modellers' project—modellers dealing with everyday energy demands—who had branched over to the Social Sciences. We could argue that being an insider in studying energy demand also gave confidence for a participant to label their work as 'sociotechnical'. This confirms our observations more generally: technical experts that are insiders to the study of domestic consumers, household technologies, and related subjects, often turn implicitly like Social Scientists, advocating the same applied methods, research questions, and even critiques. This might suggest, as Forsythe (1999) outlined in her misconceptions, that anyone can learn to be a Social Scientist over time as it is mainly common sense. But the assumption is also problematic because social scientific methods are not meant to be common sense, but to run counter to it: that is to say, they are meant to "problematize things that insiders take for granted" (Forsythe 1999, p. 130) towards which an insider is not in a privileged position.

Several other scientists in the CESI project took more distance from calling their work Social Science, but the dynamics of appropriation were still apparent: mainly in the ways of labelling what the Anthropologists and other Social Scientists in the project were expected to be doing, most typically named as, for example, 'consumer research', 'policy studies', or 'qualitative methods'. In this, what appeared was Forsythe's (1999) one key misconception that social scientific method is just about studying what people do—essentially by talking with them, whether they be consumers or policymakers—and descriptively reporting it as a result. In the Norwegian case in Chap. 3, a common expression in these large-scale collaborations was that Social Scientists are addressed in them as 'the people experts'. These kinds of labels set between disciplines may be relatively well-meaning and simply call for inclusion of other perspectives, but they can have severe consequences if we draw from Forsythe's (1999) critique of the expertise implied by them. Ethnography and other qualitative methodologies involve considerable discipline and rigour, and much of the involved expertise is highly technical, as anyone that has taken a methods class in the Social Sciences will doubtlessly know (Robison and Foulds 2019). Thus, when this expertise is underappreciated—that is to say, if the Social Sciences are just seen as a matter of common sense and talking to people—that may encourage short-term studies whose value and rigour may be questionable for the Social Scientists, but also for the research projects at large.

Finally, in the Finnish case in Chap. 4, we saw perhaps the closest expression of what Forsythe (2002, p. 133) calls "deleting" the field of the Social Sciences. In this project that studied the perceptions of laypeople on electricity reliability, qualitative accounts on this reliability were routinely described as 'subjective' and by implication as 'non-objective', 'soft', or even 'unscientific'. However, this assumption is highly dubious, once we consider the histories of the fields that study similar issues. The examination of the costs of electricity reliability is a relatively novel topic. The earliest source that our report on this topic (Silvast et al. 2006) cites was published in 1989 (Wacker and Billinton 1989), and the oldest sources cited in that early paper are in turn from 1972. It is plain that the disciplines of Anthropology, Sociology, and Social Psychology are much older than Customer Cost Studies, and that these disciplines have been routinely applied historically to the study of risk perception (e.g. Douglas and Wildavsky 1983) and to the disruption of normal routines as an impact of different disasters (e.g. Quarantelli 1954, 1960), which could also involve infrastructure disruptions. This accrued knowledge, which would have been directly relevant for examining the costs of electricity interruptions as risks or as disruptive events sociologically, was therefore not utilised because of the hierarchies and divisions of knowledge implied in this project setting.

That said, the first author in this project also experienced other kinds of appropriation, once again resembling what Forsythe (1999, 2002) outlines in her critique. While seen as 'subjective', the qualitative accounts of customer costs were clearly also relevant, and the author was routinely invited to project meetings and later, to talk in industry training events and even write a series of columns to a Finnish trade journal on these issues. These communications were never about the costs or other statistical knowledge, but mainly about how customers 'experience' power failures. Hence, while Social Science–generated evidence was almost entirely 'deleted' from the scientific report and the regulatory model that it informed, it seemed to have found another use in this context. Social Sciences became the servant of Customer Experience Studies, as it was assumed that qualitative accounts from everyday life would discover what the customers actually think and do, and this could be relevant for power companies, for example, in dealing with customer complaints or corporate communications. Such a role can have policy impact in its own right, but it may not speak to an increasing amount of interdisciplinarity assumed to be taking place.

5.6   Interpretative Flexibility

Do commentators agree on what interdisciplinarity means? This section discusses the issue and moves to a commonplace tool from STS: the idea that technologies and concepts have interpretative flexibility. Interpretative flexibility has started to refer to any flexible meanings in general, but the concept has more particular roots, which we visit briefly now to qualify how the concept has been used and informs our own interpretations herein.

The idea of interpretative flexibility has two closely related origins in the history of STS (see summary in Pinch and Bijker 1984). Firstly, within the Empirical Programme of Relativism—a branch of Science Studies—which has focused on disputed knowledge and scientific controversies. Science Studies scholars used these controversies methodologically to examine how social negotiations explain the status of some, especially disputed scientific findings. Namely, when scientists conduct experiments and discover new data, and when these cannot be explained by the established knowledge, the findings acquire interpretative flexibility: what is their correct explanation? Because the scientific findings may have more than one interpretation, social negotiations are drawn upon to close these debates, although the closure may only last for a time. It should be stressed though that the links between scientific facts and external social forces are complex and notoriously difficult to show, and the authors of this theory were not advocating the view that scientific knowledge is a simple result of social agreement. Instead, they simply argued that social processes can play a role in scientific processes, especially when disputed findings are at stake.

Secondly, STS scholars Pinch and Bijker (1984) took the idea of interpretative flexibility and applied it anew in the case of technological development. Their methodological strategy was to study controversies about technology to highlight, again, how social negotiations play into resolving them. The interpretative flexibility in this case refers not to scientific findings but to technologies as such: different social groups have different ideas about what technologies mean, how they should be designed, and how they work. These groups—called relevant social groups when they share the same interpretation of a studied technology—seek to establish their view and eventually the technological controversy is closed and stabilised. Once again, this closing of the debate may be influenced by social processes that are not necessarily scientific or technological, although how that happens is an empirical problem and not straightforward. The pertinent conclusion is that any technology—such as the oft-used example, bicycle—has no single interpretation during its inception and development, but multiple groups interpret the uses, designs, risks of technology, and so forth, differently. This then has an impact on what that technology becomes like.

In this book, we can draw on all the meanings of interpretative flexibility introduced here: the two meanings in classic STS; and the one general meaning, which simply states that concepts have flexibility when they are interpreted differently. Firstly, the Science Studies meaning introduces an intriguing angle to interpretations that happen within interdisciplinary projects. While the classic works talked about a core set of scientists that close controversial debates, in interdisciplinary projects the main actors offer a much wider scope of different kinds of expertise and perspectives. We argue that our cases—such as the energy computer models in Chap. 2, the innovative energy collaborations in Chap. 3, and the cost calculations in Chap. 4—could become more mired in controversy as more epistemic cultures begin interpreting their results. The opening to greater epistemological variety makes integrative interdisciplinary research difficult as a practical matter.

Indeed, it could be said that interdisciplinary projects even explicitly set the stage for the kinds of scientific controversies examined in Science Studies. For an energy computer modeller, for example, a modelling result may be undisputed enough (although we do not want to underestimate the complexity of the interpretation of what constitutes a 'valid' finding from computer models, see Silvast et al. 2020). Yet, Economists, Sociologists, Legal scholars, or Ethicists might have entirely different interpretations of the implications of the same results. Especially theories and methods that lie at the intersection of different disciplines—such as Engineering methods that seek to synthesise expert opinions or the cost calculations of complex risks in Chap. 4—contain potential for causing disputes when their inner logics and functioning are exposed to interpretation by multiple academic disciplines. These kinds of controversies can be generative for new ideas as such, but clearly also slow down the pursuit of science.

The technological meaning of interpretative flexibility from Pinch and Bijker (1984) also finds it corollary in interdisciplinary working. Here, the matter is not so much that findings are disputed, but that different stakeholders will have various interpretations of how energy technologies should be designed and used. A typical example is offered by wind power. Place attachment and concerns about landscape and fairness are among the many factors that are known to affect the interpretation of wind installation. Discussions and opposition have been afloat in many countries that have installed large-scale wind power (see Delicado et al. 2014).

The wind power controversy is outside of our scope, but we draw on it to highlight that interdisciplinary projects, and especially transdisciplinary projects that transcend to non-academics as co-designers of researchers, could increase the number of such disputes that need to be tended to by research. This happens mainly because more and different flexible interpretations will be made effective in these projects. For instance, we would argue that the future energy scenarios examined in Chap. 2 have been shaped by this process: because the future energy technologies that they include, or exclude, are so readily disputed by different stakeholders, it has been relatively slow to produce the scenarios. In other words, interdisciplinary methods that promise to cover a considerable amount of ground also expose a space for controversies considering the flexible interpretations about technology. As Pinch and Bijker (1984) explained, many controversies and disputes are eventually resolved, but sometimes this does not happen by consensus; at times, it can require re-interpreting what the problem was in the first place. Undoubtedly, interdisciplinary projects will need a toolbox of such strategies available if they are to succeed.

Lastly, we can draw on the more generous meaning of interpretative flexibility and note that it has resonance to interdisciplinary working. It is by now well established that the concept of interdisciplinarity has no single meaning and cannot be defined in any exactitude (Huutoniemi et al. 2010; Klein 2010). It is thus obviously clear that interdisciplinarity, multidisciplinarity, transdisciplinarity, and crossdisciplinarity are often used interchangeably, since they have interpretative flexibility. For example, while interdisciplinary and multidisciplinary are sometimes simply exchangeable, at other times, a difference is made between them that suggests academic interdisciplinarity literature was drawn upon. The flexible use of these terms is not to be dismissed as such (there are no clear alternatives that would offer a deeply anchored designation, so any conceptual taxonomies that work should be seen as adequate for their purpose). But it is still worth reminding ourselves that the term 'interdisciplinarity' conceals a considerable amount of flexible possibilities for research designs, approaches, methods, and theories that are not always explicated when the term is used.

Another relatively common and established conclusion is that interdisciplinary projects give rise to different interpretations of the core concepts that they use. As we have seen in Chap. 2, these flexible concepts include energy demand, energy scenarios, and even the very concept of what is an energy model. These terms might be understood differently in various energy-related sectors and epistemic cultures, and the same could be said of the key terms drawn upon in Chap. 3 (e.g. environmental innovation) and Chap. 4 (e.g. energy costs). Here we visit this theme only relatively briefly because there are good lexicons available that precisely address the diverse interpretations of energy-related language and what uses these concepts could imply (e.g. Foulds and Robison 2017).

Huutoniemi et al. (2010) come to other interesting conclusions when they study how interdisciplinarity manifests in research proposal writing in Finland. Here, the interpretative flexibility of interdisciplinarity is not merely an abstract observation. It is genuinely difficult to find out when interdisciplinarity is more of a rhetoric and when it concerns work that explicitly integrates academic disciplines. As we argued in Sect. 5.2 on epistemic cultures, aligning with what Huutoniemi et al. (2010) propose, the working practices of scientists offer one route to observe this integration empirically. Related to this, the degree of interdisciplinarity is often difficult to value in exactitude. As Chap. 1 outlined, interdisciplinarity simply means that more than one discipline is brought together. But then the question becomes, which disciplines should these be? When Power Systems Engineering is brought together with High Voltage Technology approaches, is this interdisciplinary? Or is it only when Sociologists are working with Power Systems Engineers? While there seemingly is no privileged answer, and both are true by strict definition of interdisciplinarity, some such as Winskel (2018) suggest distinctive forms of interdisciplinarity—called whole systems research—which work across a wide disciplinary range. Thus, there is more research that needs to be done on overcoming the interpretative flexibility of interdisciplinarity and making distinctions that scholars and practitioners can use, such as between more radical and less radical versions of interdisciplinarity. Another potential route, which will we end with, is to ask once again: why are academic disciplines important, even in highly interdisciplinary circles?

5.7   The Importance of Disciplines

This book has examined the social activities that surround interdisciplinary energy research. It has highlighted lessons, strategies, stories, and what works and does not work. While we have been highly focused on these sociological dynamics in this book, these lessons are obviously not meant to claim that interdisciplinarity does not work, whether in theory or in practice. It should be clear that we are both vocal supporters of more interdisciplinarity in energy systems and have spent most of our scholarly careers in advocating these kinds of research and study programmes.

We want to end, though, on a slightly more ambivalent note as concerns the academic literature on interdisciplinarity. Before we start this, we want to stress the value of these new concepts and their extensive discussion over the past decades. This has not only generated academic impact that has lasted for decades, but deeply affected how scholars addressing grand societal challenges now work. A good example is given by the UK, where the amount of interdisciplinary and cross-institute energy networks has grown too extensive to fit it into one review, as was mentioned in Sect. 5.1.

Yet, we must also admit to certain conceptual ambivalence about the discourse of interdisciplinarity. It is clearly the case that interdisciplinarity, transdisciplinarity, and multidisciplinarity have received numerous designations and detailed discussions both in academic works and in official documents (see Klein 2010 for a review). That said, many whole volumes and books published on interdisciplinarity contain no attempt to define what is being integrated by being interdisciplinary—namely, what is an academic discipline? To use the terms of this chapter, 'discipline' appears to have become a kind of a boundary object (Star 2010) that integrates various social worlds that talk to the increasing importance of interdisciplinarity, yet very few try to designate what exactly disciplines are.

However, there are important pockets to look for in STS for understanding disciplines, the most famous of which is physicist Thomas Kuhn's (1962) foundational work on *scientific paradigms*. STS scholar, Mike Michael (2017), has also characterised what a paradigm implies: that traditions of scientific research cohere because of a combination of techniques and theories that scientists immerse in. Paradigms grant scientists presupposed ideas and background assumptions, and grounds for training students into the scientific community via specific instruments, ideas, and practices. While paradigms are not the same as disciplines—instead, one discipline has multiple paradigms, as one is superseded by others during scientific revolutions—the important point is that interdisciplinary scholars (including those in our studies in this book) do draw on a large set of assumed instruments and practices, and often explicitly name them as disciplines.

Therefore, the interdisciplinary grey papers in Chap. 2 talk to the importance of Statistics and Mathematics, and discipline-crossing computer models are frequently rooted in the discipline of Physics; the FMEs in Chap. 3 are divided into 'technological' and 'social scientific' centres; and the cost assessors in Chap. 4 still identify themselves as Power Systems researchers. Meanwhile, the existence of energy-related SSH has become an almost universally accepted idea in Europe, not only among research funders (e.g. the Norwegian Research Council, as examined in Chap. 3), but also among interdisciplinary academics many of whom would contest the idea of a single discipline of Social Science (e.g. Foulds and Robison 2018; Royston and Foulds 2019). This is not contradictory. Even without claiming that disciplines are like monoliths that mostly exist in distant university faculty structures, there are grounds to defend the rigour of possessing certain traditions of thought and presupposed ideas that give scholars their standards of objectivity, forms of proof, and conditions for understanding quality. Without possessing any backgrounded ideas and ways of attributing meaning to findings, knowledge production would become practically impossible, perhaps a kind of never-ending controversy that STS scholars describe in the context of interpretative flexibility; a dispute without a closure (Pinch and Bijker 1984). Therefore, we argue, scholars should still call themselves Anthropologists, Sociologists, Geographers, or any other disciplinary identities they resonate with, even amid the influential discourse on increasing interdisciplinarity.

We discussed in Chap. 1 that Jacobs (2013, p. 29) likened disciplines to professions as both involve scholarly associations, conference meetings, and publication in specific peer-reviewed journals. Of course, these definitions are up for debate and we, like many other interdisciplinary Social Scientists that we know, remain ambivalent about the importance of disciplines and interdisciplines. One could certainly and easily show that various interdisciplines—such as our own STS—now host their own journals, conferences, associations, and avenues for receiving doctoral degrees, and have set up such arrangements with great efforts whilst remaining rigorously interdisciplinary by identity. The same is true for several other established, new, and emerging fields: from Gender Studies to European Studies, Transport Studies, and much else.

Therefore, both disciplines and interdisciplinarity matter: this much is clear to decades of interdisciplinarity literature. The journey that this book has taken has tried to stress the importance of reflexivity and the situated nature of disciplinary resources and constraints. All those interested in well-functioning interdisciplinary systems should remain aware of the different degrees of recognition and influence that the new interdisciplines have. There is much dynamism and breadth in new interdisciplines—although the same has been said of disciplines—but scholars and students have to confront the confines of the funding structures, publishing practices, and academic positions when pursuing interdisciplinarity in practice. The opportunities for interdisciplinary collaboration can bear fruit only when paired with recognition of academic disciplines to do this work successfully and with rigour. For all those practising interdisciplinary energy research—SSH researchers included—both ambivalences and opportunities are inescapable.

REFERENCES

Balmer, A.S., Calvert, J., Marris, C., Molyneux-Hodgson, S., Frow, E., Kearnes, M., Bulpin, K., Schyfter, P., MacKenzie, A., Martin, P., 2015. Taking roles in interdisciplinary collaborations: Reflections on working in post-ELSI spaces in the UK synthetic biology community. Science and Technology Studies 28, 3–25. https://doi.org/10.23987/sts.55340

Barry, A., Born, G., Weszkalnys, G., 2008. Logics of interdisciplinarity. Economy and Society 37, 20–49. https://doi.org/10.1080/03085140701760841

Beaulieu, A., 2010. Research note: From co-location to co-presence: Shifts in the use of ethnography for the study of knowledge. Social Studies of Science 40, 453–470. https://doi.org/10.1177/0306312709359219

Christensen, T.H., Friis, F., Bettin, S., Throndsen, W., Ornetzeder, M., Skjølsvold, T.M., Ryghaug, M., 2020. The role of competences, engagement, and devices in configuring the impact of prices in energy demand response: Findings from three smart energy pilots with households. Energy Policy 137, 111142. https://doi.org/10.1016/j.enpol.2019.111142

Clarke, A.E., Star, S.L., 2007. The social worlds framework: A theory/methods package, in: Hackett, E.J., Amsterdamska, O., Lynch, M., Wajcman, J. (Eds.), The handbook of science and technology studies. The MIT Press, Cambridge, pp. 113–137.

Davenport, S., Leitch, S., Rip, A., 2003. The 'user' in research funding negotiation processes. Sci. Public Policy 30, 239–250. https://doi.org/10.3152/147154303781780362

Delicado, A., Junqueira, L., Fonseca, S., Truninger, M., Silva, L., 2014. Not in anyone's backyard? Civil society attitudes towards wind power at the national and local levels in Portugal. Science and Technology Studies 27, 49–71. https://doi.org/10.23987/sts.55324

Douglas, M., Wildavsky, A., 1983. Risk and culture: An essay on the selection of technological and environmental dangers. University of California Press, Oakland.

Forsythe, D.E., 2002. Studying those who study us: An anthropologist in the world of artificial intelligence. Stanford University Press, Redwood City.

Forsythe, D.E., 1999. "It's just a matter of common sense": Ethnography as invisible work. Computer Supported Cooperative Work 8, 127–145. https://doi.org/10.1023/A:1008692231284

Foulds, C., Christensen, T.H., 2016. Funding pathways to a low-carbon transition. Nature Energy 1, 16087. https://doi.org/10.1038/nenergy.2016.87

Foulds, C., Robison, R., 2018. Mobilising the energy-related social sciences and humanities, in: Advancing energy policy: Lessons on the integration of social sciences and humanities. Palgrave Macmillan, Cham, pp. 1–12.

Foulds, C., Robison, R., 2017. The SHAPE ENERGY Lexicon—interpreting energy-related social sciences and humanities terminology. SHAPE ENERGY, Cambridge.

Foulds, C., Royston, S., Berker, T., Nakopoulou, E., Abram, S., Ančić, B., Arapostathis, E., Badescu, G., Bull, R., Cohen, J., Dunlop, T., Dunphy, N., Dupont, C., Fischer, C., Gram-Hanssen, K., Grandclément, C., Heiskanen, E., Labanca, N., Jeliazkova, M., Jörgens, H., Keller, M., Kern, F., Lombardi, P., Mourik, R., Ornetzeder, M., Pearson, P., Rohracher, H., Sahakian, M., Sari, R., Standal, K., Živčič, L., 2020. 100 Social Sciences and Humanities priority research questions for energy efficiency in Horizon Europe. Energy-SHIFTS, Cambridge.

Foulds, C., Jones, A., & Bharucha, Z. P., 2021. UK aid and research double accounting hits SDG projects. Nature 592(7853): 188–188. https://doi.org/10.1038/d41586-021-00895-2

Genus, A., Iskandarova, M., Goggins, G., Fahy, F., Laakso, S., 2021. Alternative energy imaginaries: Implications for energy research, policy integration and the transformation of energy systems. Energy Research and Social Science 73, 101898. https://doi.org/10.1016/j.erss.2020.101898

Geuna, A., 2001. The changing rationale for European University research funding: Are there negative unintended consequences? Journal of Economic Issues 35, 607–632. https://doi.org/10.1080/00213624.2001.11506393

Gibbons, M., 2000. Mode 2 society and the emergence of context-sensitive science. Science and Public Policy 27, 159–163. https://doi.org/10.3152/147154300781782011

Goldfarb, B., 2008. The effect of government contracting on academic research: Does the source of funding affect scientific output?. Research Policy 37, 41–58. https://doi.org/10.1016/j.respol.2007.07.011

Haddon, L., 2011. Domestication analysis, objects of study, and the centrality of technologies in everyday life. Canadian Journal of Communication 36, 311–323.

Hargreaves, T., Burgess, J., 2009.Pathways to Interdisciplinarity: A Technical Report Exploring Collaborative Interdisciplinarity Working in the Transition Pathways Consortium. University of East Anglia, Norwich.

Huutoniemi, K., Klein, J.T., Bruun, H., Hukkinen, J., 2010. Analyzing interdisciplinarity: Typology and indicators. Research Policy 39, 79–88. https://doi.org/10.1016/j.respol.2009.09.011

Impello, 2018. Effekter av energiforskningen. [The Effects of Energy Research.] Impello, Trondheim.

Jacobs, J.A., 2013. In defense of disciplines: Interdisciplinarity and specialization in the research university. University of Chicago Press, Chicago.

Kania, K., Bucksch, R., 2020. Integration of social sciences and humanities in Horizon 2020: Participants, budgets and disciplines—5th monitoring report on projects funded in 2018 under the Horizon 2020 programme. European Commission Directorate-General for Research and Innovation, Brussels.

Kasperowski, D., Hillman, T., 2018. The epistemic culture in an online citizen science project: Programs, antiprograms and epistemic subjects. Social Studies of Science 48, 564–588. https://doi.org/10.1177/0306312718778806

Klein, J.T., 2010. A taxonomy of interdisciplinarity, in: Klein, J. T., Mitcham, C., Frodeman, R. (Eds.), The oxford handbook of interdisciplinarity. Oxford University Press, Oxford, pp. 15–30.

Knorr Cetina, K., 1999. Epistemic cultures: How the sciences make knowledge. Harvard University Press, Cambridge.

Kruse, C., 2021. Attaining the stable movement of knowledge objects through the Swedish criminal justice system. Science and Technology Studies 34, 2–18. https://doi.org/10.23987/sts.80295

Kruse, C., 2015. The social life of forensic evidence. University of California Press, Oakland.

Kuhn, T.S., 1962. The structure of scientific revolutions. University of Chicago Press, Chicago.

Longhurst, N., Chilvers, J., 2012. Interdisciplinarity in transition? A technical report on the interdisciplinarity of the transition pathways to a low carbon economy consortium. Science, Society and Sustainability, University of East Anglia, Norwich.

Mallaband, B., Wood, G., Buchanan, K., Staddon, S., Mogles, N.M., Gabe-Thomas, E., 2017. The reality of cross-disciplinary energy research in the United Kingdom: A social science perspective. Energy Research and Social Science 25, 9–18. https://doi.org/10.1016/j.erss.2016.11.001

Manjarrés-Henríquez, L., Gutiérrez-Gracia, A., Vega-Jurado, J., 2008. Coexistence of university-industry relations and academic research: Barrier to or incentive for scientific productivity. Scientometrics 76, 561–576. https://doi.org/10.1007/s11192-007-1877-7

Michael, Mike, 2017. Actor-network theory: Trials, trails and translations. Sage, London.

Nowotny, H., Scott, P.B., Gibbons, M.T., 2001. Re-thinking science: Knowledge and the public in an age of uncertainty. Polity Press, Cambridge.

Overland, I., Sovacool, B.K., 2020. The misallocation of climate research funding. Energy Research and Social Science 62, 101349. https://doi.org/10.1016/j.erss.2019.101349

Philipps, A., Weißenborn, L., 2019. Unconventional ideas conventionally arranged: A study of grant proposals for exceptional research. Social Studies of Science 49, 884–897. https://doi.org/10.1177/0306312719857156

Pinch, T.J., Bijker, W.E., 1984. The social construction of facts and artefacts: Or how the sociology of science and the sociology of technology might benefit each other. Social Studies of Science 14, 399–441. https://doi.org/10.1177/030631284014003004

Quarantelli, E.L., 1960. Images of withdrawal behavior in disasters: Some basic misconceptions. Social Problems 8, 68–79. https://doi.org/10.2307/798631

Quarantelli, E.L., 1954. The nature and conditions of panic. The American Journal of Sociology 60, 267–275. https://doi.org/10.1086/221536

Robison, R., Foulds, C., 2019. 7 principles for energy-SSH in Horizon Europe: SHAPE ENERGY Research & Innovation Agenda 2020–2030. SHAPE ENERGY, Cambridge.

Royston, S., Foulds, C., 2019. Use of evidence in energy policy: The roles, capacities and expectations of Social Sciences and Humanities: Scoping workshop report. Energy-SHIFTS, Cambridge.

Royston, S., & Foulds, C., 2021. The making of energy evidence: How exclusions of Social Sciences and Humanities are reproduced (and what researchers can do about it). Energy Research & Social Science, 77, 102084. https://doi.org/10.1016/j.erss.2021.102084

Salmenkaita, J.-P., Salo, A., 2002. Rationales for government intervention in the commercialization of new technologies. Technology Analysis & Strategic Management 14, 183–200. https://doi.org/10.1080/09537320220133857

Schot, J., Steinmueller, W.E., 2018. Three frames for innovation policy: R&D, systems of innovation and transformative change. Research Policy 47, 1554–1567. https://doi.org/10.1016/j.respol.2018.08.011

Silvast, A., 2020. UKERC energy research Atlas: Interdisciplinary whole systems research. UK Energy Research Centre, London.

Silvast, A., Heine, P., Lehtonen, M., Kivikko, K., Mäkinen, A., Järventausta, P., 2006. Sähkönjakelun keskeytyksestä aiheutuva haitta. Espoo.

Silvast, A., Laes, E., Abram, S., Bombaerts, G., 2020. What do energy modellers know? An ethnography of epistemic values and knowledge models. Energy Research and Social Science 66, 101495. https://doi.org/10.1016/j.erss.2020.101495

Silvast, A., Virtanen, M.J., 2019. An assemblage of framings and tamings: Multi-sited analysis of infrastructures as a methodology. Journal of Cultural Economics 12, 461–477. https://doi.org/10.1080/17530350.2019.1646156

Star, S.L., 2010. This is not a boundary object: Reflections on the origin of a concept. Science, Technology & Human Values 35, 601–617. https://doi.org/10.1177/0162243910377624

Taylor, P.G., Upham, P., McDowall, W., Christopherson, D., 2014. Energy model, boundary object and societal lens: 35 years of the MARKAL model in the UK. Energy Research and Social Science 4, 32–41. https://doi.org/10.1016/j.erss.2014.08.007

UCL, 2021. UK MARKAL [WWW Document]. URL https://www.ucl.ac.uk/energy-models/models/uk-markal (accessed 3.23.21).

Wacker, G., Billinton, R., 1989. Customer cost of electric service interruptions. Proceedings of the IEEE 77, 919–930. https://doi.org/10.1109/5.29332

Wang, X., 2009. UKERC energy research landscape: Interdisciplinary Research Centres. UK Energy Research Centre, London.

Winskel, M., 2018. The pursuit of interdisciplinary whole systems energy research: Insights from the UK Energy Research Centre. Energy Research and Social Science 37, 74–84. https://doi.org/10.1016/j.erss.2017.09.012

INDEX[1]

[1] Note: Page numbers followed by 'n' refer to notes.

© The Author(s) 2022   121
A. Silvast, C. Foulds, *Sociology of Interdisciplinarity*,
https://doi.org/10.1007/978-3-030-88455-0